# GENES, FLIES, BOMBS AND A BETTER LIFE

## Pitchpole Books

# GENES, FLIES, BOMBS AND A BETTER LIFE

## in the footsteps of Hermann Muller

Geoff Meggitt

Published 2016 by Pitchpole Books
Cheshire, England
pitchpole.co.uk

Copyright © Geoff Meggitt 2016

All rights reserved. No part of this book may be reproduced or transmitted in any form or by any means, electronic and mechanical, including photocopying, recording, or by any information storage and retrieval system, without permission in writing from the Publisher.

*ISBN: 978-0-9575549-7-9*

*To Joyce*

# ACKNOWLEDGEMENTS

Images are credited in their captions. Where there is no credit given the image is believed to be in the public domain or it has not been possible to trace the copyright holder. My apologies for any failures to acknowledge ownership of material or copyright.

My thanks to the staff of the Lilly Library at the University of Indiana, the Cold Spring Harbor archives, the Wellcome Library in London and the University of Manchester Library. I would also like to thank William de Jong Lambert for conversations that opened up new areas of Muller's life to me.

## Table of Contents

PREFACE..................................................................ix
1  GENESIS....................................................................1
2  EVOLUTION, CELLS AND CHROMOSOMES .............20
3  FLIES, GENES AND MUTATIONS................................41
4  MUTATIONS FLOURISH.................................................53
5  EVOLUTION AND EXILE ...............................................70
6  BEFORE THE BOMB.....................................................102
7  THE BOMB....................................................................114
8  FALLOUT.......................................................................121
9  RADIATION HARMS....................................................133
10  A BETTER SPECIES?....................................................179
11  A CHEMICAL GENETICS............................................223
12  POSTSCRIPT..................................................................230
13  ENDNOTES....................................................................234
14  GLOSSARY.....................................................................252
15  BIBLIOGRAPHY............................................................259
16  INDEX.............................................................................283

## Illustration Index

1: Replica of a Leeuwenhoek microscope................................3
2: Dissection of a cheese skipper by Swammerdam................5
3: The Chain of Being................................................................10
4: Drosophila melanogaster ......................................................44
5: Muller and Altenburg 1922..................................................57
6: The Spark cartoon..................................................................64
7: Crossroads Baker..................................................................125
8: An Ulster County Family ...................................................197
9: The Eugenics Tree 1921.......................................................204
10: Muller in 1956.....................................................................230

# PREFACE

This book started with a simple enough puzzle. I wondered why, in the 1970s, there had been a dramatic turnaround in the perception of the hazards of nuclear radiation. Before that time it seemed all about genetic effects; after it, the cancer hazards became more important and quickly dominated the field.

I already knew of the pioneering work of Hermann Muller: he had discovered that radiation caused mutations in fruit flies in the 1920s. However I hardly realised that this had been a major stimulus to basic research in genetics and was key to the first attempts at mapping the chromosomes. I hadn't realised too that this was so close to the time when it became apparent that chromosomes were the genetic material. So close too to the rediscovery of Mendel's work that set the foundations for modern genetics.

It became imperative to find out more about what passed for genetics before Mendel – how could I understand his importance otherwise? So that took me far back, further than I intended, and the first chapter is a loose account of early ideas on heredity.

Then I read some of Muller's story. This remarkable and difficult man had been both a gad-fly and central to the story of genetics from the recognition of the role of the chromosome to the discovery of DNA and molecular biology in general. His story seemed to touch on not just technical genetic matters, at which he was a master, but on social issues like attitudes to Communism, eugenics, the hazards of radiation, the corruption of science and, generally, the role of the scientist in public life. He had seen some of the 20$^{th}$ century's horrors: escaping the two evil poles of Nazi Germany and Stalin's Russia and suffering for his socialist convictions for a long time after. He wanted to do good, to improve man's lot – and man himself.

The book became the challenge of weaving together these threads of Muller's life, his achievements and those of many others into something readable, balanced and informative. The reader must judge. For me, I managed to answer my question.

# 1 GENESIS

Must we ascribe every effect of the cause whereof wee are ignorant to some cause produced *de novo* and immediately from God?

William Boghurst 1666[1]

## GENERATION

The notion of inheritance – that a child resembles its parents or grandparents – must be ancient. In the *Iliad*, written about the 8th century BC from earlier oral tradition, it was taken for granted that the son inherits the father's heroic character. How it happened was a matter for speculation. Hippocrates (ca 460-377 BC) thought that the "seed material" somehow gathered from all the body into the reproductive organs of the mother and father and came together in fertilisation. The theory could accommodate the fact the certain characteristics were passed on ("he has his father's eyes but his mother's mouth") but it allowed (perhaps encouraged) the idea that acquired characteristics could be inherited. The idea survived as "pangenesis" for over 2000 years and was essentially the explanation favoured by Charles Darwin.

## EPIGENESIS

Aristotle (384-322BC), like others before him, had studied the development of chick embryos[2] and thought that organisms arose from a homogeneous mass and that the individual organs emerged one after another: a process called epigenesis. The process might be initiated by spontaneous generation (for the lower animals), asexual or sexual reproduction. For humans and other mammals the embryo was made from the matter of the menstrual blood, shaped by the male semen and nourished by blood from the female. This was a plausible idea at the time, given the tools and information he had, explaining the known facts. It was deficient in being wrong but it, or something very like it, remained the accepted wisdom for around two millenia until discredited by the work of William Harvey (1578-1657). In terms of inheritance it had one essential element (at least in some of Aristotle's work) that was important: it recognised that a foetus was a combination of its mother and father – with perhaps characteristics of its grandparents too.[3]

There were, of course, other Greeks but this is not the place to discuss them. It is enough to quote Ernst Mayr [4] "... the major contribution made by the Greeks was that they introduced an entirely new attitude towards inheritance. They considered it no longer as something mysterious, given by the gods, but as something to be studied and to be thought about. In other words they claimed inheritance for science."

At the beginning of the 17th century there were two theories of generation: epigenesis and preformation but the dominant one, having been favoured by Aristotle, was epigenesis.

William Harvey is best-remembered for his discovery of the circulation of the blood published in 1628 but later in his life he spent several years studying generation, trying to obtain empirical support for Aristotle's ideas. Appointment as "Physician in Ordinary" to Charles I made the corpses of female deer hunted in the royal park at Richmond in London available for dissection. He found what he thought were eggs. They were in fact undeveloped embryos but the discovery did lead him to propose that all animals came from eggs. Perhaps more importantly he was unable to find any evidence of the Aristotelian version of generation in mammals. However his observations on chicken eggs did convince him that generation came by epigenesis so, in that, he was a supporter of Aristotle. His book *On the Generation of Animals* published in 1651 was perhaps the last major work in biology that did not owe something to the microscope. It was this instrument that was to be responsible for major advances and significant confusion for the next hundred years.

## MICROSCOPES [5]

It is generally accepted that a crude compound microscope with two lenses was invented by Sacharias Jansen (spellings vary slightly), a Dutch spectacle maker, in about 1590. Magnification was quite low (perhaps three to ten times) and the quality of the image was no doubt poor. The first significant scientific use of the compound device was by Marcello Malpighi (1628-1694) in his pioneering work of the mid-1600s that established the science of micro-anatomy. The first person to see blood capillaries (and thus provide the missing link between the arteries and veins in Harvey's circulation scheme), he studied developing chicken eggs and the anatomy of insects, notably the silkworm in 1661. However, the book that created the most general interest was Robert Hooke's remarkable *Micrographia* published in 1665. Hooke (1635-1703) used a compound microscope and the book contains his superb drawings of many tiny things, of which the most striking are perhaps the flea and the louse.

# GENESIS

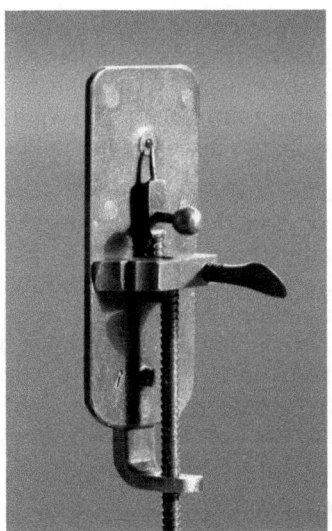

Illustration 1: Replica of a Leeuwenhoek microscope

(J Rouwkema)

These early studies had a general impact on the generation debate: animal structures, however small, seemed to have no granularity and there seemed no limit to how small they could be. There was, although Hooke had seen plant cells in cork, no sign of the animal cell.

The early compound microscope suffered from spherical and chromatic aberration and as a result had a rather limited magnifying power, say 10-20x. Simple single-lens devices had been in use for several hundred years and were to be better instruments at high power than the compound microscope for some time to come. Indeed it was well into the 19th century before the use of compound lenses solved the problem of chromatic aberration allowing the full potential of the compound microscope to be exploited. Some of the best of the single-lens devices were made by the the Dutchman Anthony van Leeuwenhoek (1632-1723) who made powerful devices (with magnifications up to nearly 300X and maybe to 500X) by using one tiny bi-convex lens. These extraordinary little lenses – just a few millimetres in diameter at most – seem to have been made from globules of melted glass finished by grinding and polishing (there is doubt about exactly how he did it because he was secretive about his methods) and he is believed to have produced more than 500 microscopes from them.

# GENES, FLIES, BOMBS...

To make the microscopes a lens was held against a small hole in a sheet of metal and held in place by riveting a similar piece on top. A number of screws allowed a specimen (impaled on a small spike) to be moved around the field of view and brought into focus. The whole microscope was about 80 mm long and was held close up to the eye to view.

Just nine are known to have survived; 26 examples presented to the Royal Society disappeared in the 19th century. With these instruments (which produced magnifications of nearly ¼ that available from optical microscopes today[6]), and perhaps inspired by *Micrographia,* Leeuwenhoek made a number of discoveries from around 1670 including rotifers (microscopic animals that inhabit water), bacteria and was one of the first to see spermatozoa in 1677 in a sample brought to him by Johan Ham von Arnhem, a young student.[7] His discoveries were communicated by letter (200 of them) to the Royal Society from 1673 and were published in their *Transactions*.

It is sometimes assumed that Leeuwenhoek invented the microscope, or at least the single-lens device. In fact one of the best descriptions of how to make tiny lenses comes from the Preface to Hooke's *Micrographia*:

> ...And hence it is, that if you take a very clear piece of a broken Venice Glass, and in a Lamp draw it out into very small hairs or threads, then holding the ends of these threads in the flame, till they melt and run into a small round Globul, or drop, which will hang at the end of the thread; and if further you stick several of these upon the end of a stick with a little sealing Wax, so as that the threads stand upwards, and then on a Whetstone first grind off a good part of them, and afterward on a smooth Metal plate, with a little Tripoly, rub them till they come to be very smooth; if one of these be fixt with a little soft Wax against a small needle hole, prick'd through a thin Plate of Brass, Lead, Pewter, or any other Metal, and an Object, plac'd very near, be look'd at through it, it will both magnifie and make some Objects more distinct then any of the great Microscopes. But because these, though exceeding easily made, are yet very troublesome to be us'd, because of their smallness, and the nearness of the Object.

It is quite possible that *Micrographia* stimulated Jan Swammerdam (1637-1680), who had been observing and dissecting insects for some time, to use a microscope. He used a single-lens type like that of Leeuwenhoek as well as a compound instrument to study them and, after seeing Malpighi's *De Bombyce*, which contained a description of the dissection of the silk worm[8] larva he carried out a remarkable series of dissections of insects through the 1660s and 70s including a dissection of the digestive system of the cheese (or ham) skipper, the larva of a small fly *Piophila casei*.

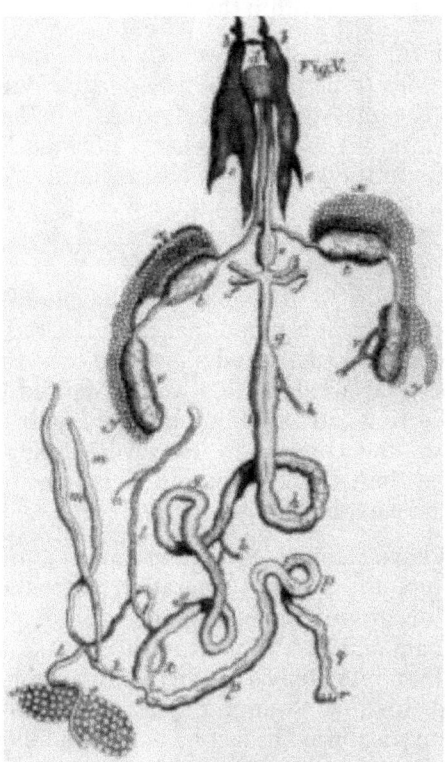

Illustration 2: Dissection of a cheese skipper by Swammerdam

This creature, about 8mm long, feasts on cheese, ripe meats and well-rotted cadavers. It gets its name from the mature larva's ability to leap 10 or more centimetres into the air by grasping its tail in its mouth, tensing its body and letting go. Able to survive passage through the human gut, where it can cause grief by burrowing through the intestine wall, it is a leading cause of myiasis, the infestation of living bodies by maggots. The dissections were extraordinary achievements in themselves with what now seem primitive tools but Swammerdam's most important contribution was probably in generation and development. He dismissed spontaneous generation and showed, in public dissections, that the various stages of insect development (egg, larva, pupa and adult) were just different forms of the same individual. He showed that the adult structures – legs, wings antennae – in fact appeared within the caterpillar before pupation.

His most significant publications were *Historia Insectorum Generalis* (*Histoire Generale des Insectes*) of 1682 , *Ephemeri Vita* 1675 (a study of the may-fly) and the massive collected works of *The Book of Nature* (Dutch *Bybel de Natuure* translated to Latin as *Biblia Natura*) completed just before his death in 1680 but not published until 1737/8 (in Dutch and Latin) and 1758 (in English).

## PREFORMATION

The alternative notion to the epigenesis favoured by Aristotle and Harvey was preformation. It claimed that the adult was already somehow pre-formed in one of its parents and simply had to grow. Leeuwenhoek seems to have been accused of this and distanced himself from the concept[9] but he did find "an exceeding minute Lamb" in the uterus of a ewe just three days after conception. However, Leeuwenhoek was much more of an observer than a speculator and he is remembered much more for what he saw than what he thought.

Many historians have claimed Swammerdam as a preformationist. He did write: "In nature there is no generation but only propagation, the growth of parts. Thus original sin is explained, for all men were contained in the organs of Adam and Eve." This may have been inspired by religious conviction rather than intended as a statement in favour of preformation. Cobb[10] has argued this and Swammerdam was certainly religious in a rather vague and mystical way. In fact he became a follower of the French mystic Antoinette Bourignon, gave up scientific investigation and spent some months on her island off the coast of Schleswig (Nordstrand – now a peninsula) before returning to Amsterdam and his work.

If Swammerdam's support for preformation was uncertain that of Nicholas de Malebranche (1638-1715) was not. Malebranche, a French priest and philosopher, deduced from the dissections that within each embryo was the embryo of its offspring in a series like Russian Babushka dolls. We all derived directly from Adam and Eve. For the ovist preformationists like Malebranche the tiny models were located in the egg waiting to be activated by the "generative principle", "breath" or "contagion" released by semen – the *aura seminalis*. However the microscope had revealed to Leeuwenhoek that the semen was full of minute *animalcules* or spermatozoa (little animals of the sperm). While some concluded that they were in fact parasites, Leeuwenhoek was convinced that they were part of male development and thought he saw two kinds which were obviously destined to make males and females. They were then candidates to be the carriers of the tiny models in the theory of animalculism (or spermism) which found its defining image in the work of Nicklass (Nicolaas) Hartsoeker (1656-1725) who in 1694 produced a sketch of a minute person (a homunculus) crouching in the head of a sperm.

He is often supposed to have claimed to have seen the creature but it seems he was describing what he thought we might see: "if we could see the little animal through the skin which hides it we might possibly see it as it is represented in the illustration." A supposed representation was published in 1699 by Dalenpatius[11] who was subsequently unmasked as François de Plantades, the Secretary of the Montpellier Academy. It seems quite possible that this was a hoax[12] but the spermist theory and the existence of homunculi was supported by many others who studied sperm under the poor microscopes of the time. The homunculus notion (sometimes called *emboîtement*) lingered on until the middle of the 18th century. However, spermism always posed a tricky theological problem: would God have created so many millions of tiny mannikins when only a few could survive? It seemed such a waste.[13]

Preformation, in its simple form, had also somehow to address the obvious fact that both parents contributed towards offspring: resemblances to mother or father must have been observed and discussed since man could speak. Yet if the offspring were perfectly preformed in one or the other, how could this be?

> Each seed includes a plant: that plant, again,
>
> Has other seeds, which other plants contain:
>
> Those other plants have all their seeds; and those,
>
> More plants,again, successively inclose.
>
> \* \* \* \* \* \* \* \*
>
> So Adam's loins contain'd his large posterity,
>
> All people that have been, and all that e'er shall be.
>
> Amazing thought! what mortal can conceive
>
> Such wond'rous smallness!
>
> Yet we must believe

# GENES, FLIES, BOMBS...

> What reason tells: for reason's piercing eye
> Discerns those truths our senses can't descry.[14]

## REGENERATION

Another of the challenges the preformation idea – whether spermist or ovist – had to survive was that posed by regeneration. The facts that lobsters regrew amputated claws and lizards replaced lost tails were well-known. Regeneration shook the conviction of some preformationists because there was clearly more to this than simply recreating a whole animal. The separated claw or tail died, suggesting that whatever was creating a new one was in the body of the creature. This was enough for Hartsoeker; he abandoned *emboîtement* in the 1720s.

However the work on regeneration of *Hydra* by the Swiss naturalist Abraham Trembley (1710-1784) in the 1740s was even more difficult to explain. He found that this tiny creature that lived on the underside of pond plants could regenerate itself from any part cut off. For this to work on the *emboîtement* theory, the minute models that formed the template had to be distributed throughout its body. A modification was needed and a version was proposed by the polymath René Antoine Ferchault de Réamur (1683-1757) – who had already been troubled by the challenge of biparental inheritance to the model – which saw invisible "germs" scattered around creatures' bodies ready to rebuild the organ (or part) when lopped off.[15] This and modified versions of it were to be the basis for preformation thinking for some time.

Of the ovists two of the more prominent were Charles Bonnet (1720-1793) and Lazzaro Spallanzani (1729-1799). Bonnet started out as a supporter of epigenesis but followed up Leeuwenhoek's observation that aphids could reproduce by pathenogenesis and managed, in the 1740s, to produce 30 generations of them without male intervention. Obviously each female aphid contained the "germs" of all the aphids she would produce and something similar would be true of all other species: the females contained in their ovaries models of all future individuals they and their descendants would produce. The male, if available and necessary, triggered the growth process by providing the initial nutrition.

Spallanzani worked on amphibians in the 1770s and 80s and found that frogs' eggs began to grow before they left the females' body. He interpreted this to mean that the tadpole was already present, curled up within the egg – hence ovism. He did not ignore the sperm, showing that if the seminal fluid produced during mating was caught by tight taffeta trousers fitted to the males, no fertilisation occurred. If he painted the fluid onto previously unfertilised eggs, some of them developed into tadpoles. It was a convincing demonstration that fertilisation was a physical process with no need for *aura seminalis*. The cleverly conceived and executed experiments were marred by one error: Spallanzani concluded that fertilisation was caused by the seminal fluid and not spermatozoa – which he considered were just parasites – because when he painted the eggs with semen in which the sperm had been killed some of them developed. It was 1910 before it was found that frog's eggs are liable to parthenogenic development when scratched and this may have happened when he brushed them.

The struggle between preformation and epigenesis continued. Benson[16] quotes Diderot's *Encyclopédie* of 1765 on generation: "It is, at present, a mystery for us into which we are so little advanced that the manifold attempts at explanations only serves to convince us more and more of their futililty." There is also a similarly pessimistic statement in the *Encyclopaedia Britannica* of 1771.

## CHANGE IN PREFORMATION

By the beginning of the nineteenth century the nature of the preformation theory had changed. It was no longer the little person in the sperm or egg waiting simply to grow but some sort of blueprint – Bonnet followed Réamur in using the term "germ" – which would form the basis for making the descendent.[17]

The concept had accommodated the experimental data thus far, reflected the desire of many to have a more-or-less mechanical interpretation of the world and was consistent with the idea of a Creation. We can see the reflection of our modern theories in it. From the zenith of the little people it had become a weaker theory; retreating as microscopy and thinking advanced.

Epigenesis from time to time gained ground. Its non-mechanical approach meant that it said very little that could be disproved. Indeed its strength was that, in retrospect, it said very little at all; the organisation of the organism arose from undifferentiated material by the action of some unknown organising principle. As Olby[18] has pointed out, it really said as much as could be said in the contemporary state of knowledge. As a description of what seemed to happen in generation (and particularly in the development of the embryo) it was near-perfect. However, its explanatory potential left something to be desired. It was the improvements in compound microscopes and techniques to go with them that were to be the next steps in understanding generation but the more significant advances in understanding heredity were to come from a quite different direction.

## CHAIN OF BEING

Illustration 3: The Chain of Being

The Chain of Being was an idea originating with the Greeks and adapted by Christian theologians.[19] It formed the background to most biological thinking to the middle of the 18th century. The Chain started, at the top, with God and then the angels, followed by Man at the summit of the animals, then plants and, at the bottom, minerals[20]. In more refined versions there was a careful gradation from Man to the animals via negroes and wild children to apes  It had two principles: plenitude and continuity. Plenitude required that, the Creator being perfect, he had created everything that could be created – so that all possible things already existed. Continuity was a requirement that all these things were linked together; a perfect plenitude could have no gaps. Trembley's hydra was a welcome missing link between animals and plants when it was discovered. The Chain posed a few theological problems. For example, it seemed to remove God's freedom to do whatever he wanted if He was constrained to create everything possible. Such apparent contradictions could be and were removed by clever words and the real challenge to the idea came from the natural world, and particularly from the notion of species.

## SPECIES

The existence of species (animals that could successfully interbreed) challenged the idea of continuity – and hence the Chain of Being – because it implied discontinuities or gaps in creation. Clearly, also, species had to be unchanging and fixed because everything had been perfect since Creation. The great French naturalist George Louis Leclerc de Buffon (1707-1788) solved the problems in 1748 by denying the reality of species and insisting that it was individuals that mattered. By 1765 he had changed his mind: now the species was real and a challenge to the Chain. This change began to sweep though all of biological thought from the middle of the 18th century and one of the triggers was the problems posed by hybrids, most immediately because they seemed to challenge the idea of fixity of species, an important issue too in the later Darwin controversies.

Buffon compromised in the preformation debate with the idea that there was an indivisible life material (molècules organiques) that was shaped by moulds (*moules intèrieure*) into different organisms. In his *Histoire Naturelle* of 1749 he thought males determined extremities (head, tail, limbs) of offspring while females were responsible for the internal parts and overall shape and size. He quoted reported results of seven crosses (female first); ass x horse, mare x ass, wolf x mastiff, canary x goldfinch, siskin x linnet and ewe x goat but only conducted the ewe x goat cross himself. These were slow and sometimes difficult crosses (the wolf killed the dog and savaged her keeper so badly she was put down). Quicker results would have been obtained with mice and rats but the idea of breeding vermin was repugnant and, as Olby says, "Thus did good taste restrict the choice of experimental materials."[21]

## PLANT SEXUALITY

Plants could have been used but the fact of their sexuality was established only around 1700 by Rudolf Jacob Camerarius (1665-1721) in Tübingen and it took some time to be generally accepted. Plant sexuality had been suspected earlier by the English botanists Nehemiah Grew (1641-1712) and John Ray (1627-1705) but Camerarius performed a conclusive series of experiments with several types of plant (mulberry, castor-oil plant and maize among them) in which no pollen was allowed to come near the female organs. He did this by keeping male and female plants apart (where the sexes were separate) or by removing the stamens where plants carried both sexes. In all cases, without pollen, no seed was set.

The idea was contrary to the Chain of Being because plants were supposed to be the "insensitive and asexual" base – disregarding the even less responsive minerals. There was also shock at the abundance of pollen and the relatively few "seed chambers". However Carl Linnaeus (1707-1778) classified plants on the basis of sexual organs – numbers of pistils and stamens – in his *Systema Naturae*[22] of 1735. At the time Linnaeus's idea caused outrage. Johann Siegesbeck, Demonstrator of the Botanical Gardens at St Petersburg from 1735, was shocked by both plant sex (he called it "shameful whoredom" because of the high ratio of male organs to female ones) and Linnaeus's use of it. He became a tenacious critic of Linnaeus who, according to some sources, had his revenge by naming an insignificant and useless weed genus *Siegesbeckia*.[23]

The binomial system he developed (and described in his *Systema Plantarum* of 1753 and the 10th edition of his *Systema Naturae* in 1758), in which an organism is classified with its genus and its specific name or epithet, much simplified identification and so revolutionised botany and zoology. The methods of taxonomy have changed and later systems for classifying plants took account of more factors than numbers of sexual parts[24] but the binomial classification system remains in use today.

## HYBRIDS[25]

The continuing controversy led the Imperial Academy of Saint Petersburg in 1759 to offer a prize for an essay on sexuality in plants. The prize was won by Linnaeus in the following year. In the course of defending his corner on sexuality (which left numerous doubters)[26] Linnaeus argued that, in animal hybrids, the outer layer derived from father, the inner layer including nervous system from mother. It was similar for plants: leaves, rind of stem from father, central part of flower and pith of stem from mother. This was based on observation of many apparent hybrids – although Linnaeus was rather relaxed in what he called a hybrid. It sometimes seems that, if he had not seen the plant before, he decided it had to be a hybrid of plants he had already seen.

There were just two known artificial hybrids at Uppsala: hybrid Goat's Beard produced by hand pollination and a hybrid of Speedwell, although there are some doubts about the second.[27]

He started by believing in the fixity of species but the hybrid experiments, sketchy as they seem, led him to accept that new species were formed from hybrids.

### KOELREUTER

It was the German Joseph Koelreuter (1733-1806) who was to increase the number of controlled artificial hybrids quite dramatically: he carried out over 500 different hybridisations involving 138 species and examined pollen grains from more than 1000 species. He knew of Linnaeus's hybrids but thought that there must be some natural mechanism preventing hybridisation creating new species willy-nilly. Otherwise the harmony that was supposed to have existed since the Garden of Eden would be destroyed. His motivation in his own work was to find out was this stabilising mechanism was.

His first hybrids in 1760 were of *Nicotiana* (*paniculata* x *rustica*) which grew well but completely failed to set seeds: the hybrid pollen grains were just empty husks incapable of pollination. He seemed to have found a mechanism that kept the species in order. When he crossed other species (pinks, carnations and Sweet Williams) he found much higher $F_1$ hybrid fertility and could thus grow good numbers of $F_2$ hybrids by self-pollinating. He also produced back-crosses of the $F_1$ hybrids by pollinating them with the original pure species. With such a collection he could see that the $F_1$ hybrids from a particular cross were all alike and in many characteristics intermediate between parents. $F_2$ hybrids and back crosses, on the other hand, were all different and more like one or other of the original species than their parents. So, even when hybrids were viable they produced such variety in the $F_2$ generation that it was clear that they would never breed true as new species. Indeed, since the $F_2$s were more like one or other the original species than their own parents, he could assert that this reversion of the offspring would eventually result in a return to the two original species.

Koelreuter published his early results in a small book in 1761 with supplements in 1763, 1764 and 1766 where he described 65 experiments. His later work was published in the journals of the St Petersburg Academy of Science where he worked unofficially on plant hybridisation, growing his plants in pots or in the gardens of friends. In 1763 he was appointed to a professorship at Karlsruhe and made Director of the Margrave's gardens. He appeared to have serious experimental resources at his disposal but the head gardener, Saul, had little time for the experiments and they were negelcted and often came to nothing. Saul seems to have undermined Koelreuter's position and he lost his post as Director in 1783 when his patron died. Although he kept his professorship he never managed to do the work he planned.

His results did argue against Linnaeus's ideas of species formation by hybridisation (although the mechanism made a return in the next century) but it also destroyed the preformation theory. Buffon and Linnaeus had proposed the two-layer theory of heredity as an alternative but Koelreuter managed to refute this too based on: the intermediate nature of hybrids, the identity of reciprocal crosses and the eventual return of offspring to original species.

One might have imagined that, even if there were disagreements about heredity and speciation after Koelreuter, the sexuality of plants would have been agreed. Not so. Major doubters were Franz Joseph Schelver (1778-1832) and his student August Henschel (1790-1856), a GP and university tutor in Breslau. Henschel's objection to Koelreuter's work was summed up by Olby[28]:

> If flowers are castrated and dusted with pollen from a plant of another species, you cannot expect them to behave normally.

When you add the fact that the plants were grown under artificial conditions to such violations it was small wonder that strange "hybrids" were obtained.. Henschel preferred the more spiritual and holistic approach of the German "Naturephilosophers" led by Hegel and others. Pollen, rather than an essential element of reproduction was just a sign of the release of the spiritual nature of the plant to prepare it for seed setting. Henschel set out his views in a book in 1820. It was dismissed by Julius von Sachs[29] as "interesting from the pathological rather than from the historical point of view."

Others continued to have doubts about plant sexuality, on rather more rational grounds, until the 1830s. This partly prompted the next major step: the work of Carl Friedrich von Gaertner (1732-1791).

## *GAERTNER*

The son of Koelreuter's friend and celebrated botanist Joseph Gaertner, Carl won the prize offered in 1830 by the Dutch Academy of Sciences for an answer to a question about artificial hybridisation and its potential benefits. Gaertner eventually won the prize in 1837. The winning essay, when revised and enlarged for publication in German in 1849, carried accounts of almost 10,000 separate experiments on 700 species with more

# GENESIS

than 250 different hybrids. He rejected the idea that new species could arise from wild hybridisation. Highly regarded by Mendel for the dedication and the volume of experimentation but criticised for the lack of detail in the reports, the work, Mendel thought, added little to theoretical understanding of heredity. Perhaps its major contribution was in helping Mendel define his own experimental procedures more tightly. For example, he choose easily-counted distinct characters and made sure there were enough plants in his studies to draw statistically meaningful conclusions.

## NAUDIN

French botanist Charles Naudin (1815-1899) carried out hybridisation for taxonomic studies of potatoes and cucumbers in the 1840s but became interested in the evolutionary role of hybridisation. He saw reversion of hybrid *Primula* more or less completely to the two parent species and followed this up in a series of experiments between 1852 and 1861 which were published in part in the early 1860s, and fully in 1865. He used 60 different species: 16 from the cucumber family, 11 species of *Nicotiana* and six of *Datura*. He extended the results of Koelreuter and Gaertner in one important respect: he grew on enough plants from each cross to have a good chance of seeing all the possible hybrid forms. The *Datura* results showed reversion of the F2 generation to the parent species. However he did not analyse his results to determine ratios of the three F2 types he found. He was also looking at the plants as a whole rather than for transmission of specific characteristics. His interpretation was that a hybrid was a mosaic made up of the distinct essences of the two parents. The mosaic could be so finely grained that it could no be resolved but sometimes whole organs might come from one parent or the other. His results for *Datura* were compromised by an unsuspected viral infection (which was not recognised for more than 50 years) and the difficulty in getting consistently good germination of the half-hardy plant in Paris. While he missed the insight that was to come from Mendel, he did convince himself that hybridisation could not lead to new species: reversion would see to that.

## MENDEL

Johann Mendel (1822-1884) was born in Hynčice in the Czech Republic (then Heidendorf in the Austrian Empire), the only son of a peasant farmer. A successful diligent school student he studied for two years at the Philosophical Institute of Olomouc but suffered from what seems to have been a stress-related illness perhaps brought on by his difficult financial circumstances at the time. On the recommendation of one of his tutors and to free himself from what he later called "the bitter struggle for existence" he entered the Augustinian Abbey of St Thomas at Brünn (now Brno in the Czech Republic) in 1843 taking the forename Gregor. He was found unsuited to pastoral work – visiting the sick made him ill – but the Abbey was a thriving centre for teaching and he became a successful teacher at the Gymnasium in what is now Znojmo. Although successful he was unqualified and to become a permanent teacher he had to pass teaching examinations: when he took them in 1850 he failed[30]. As a result it was recommended that he spend some time at the University of Vienna to improve his knowledge of natural science. He continued teaching, now in a temporary post at the technical high school, but spent four terms between October 1851 and August 1853 in the philosophical faculty at Vienna. There he was taught physics by Christian Doppler (who had proposed his Effect[31] a decade earlier and died during Mendel's time in Vienna) but the greatest influence on him came from the botanist Franz Unger (1800-1870). Unger, who started as a lawyer, became a physician and made contributions to palaeontology, was professor of plant physiology at the time. He had already been attacked by the Church Establishment for his evolutionary views in 1852 and narrowly escaped suspension. In his textbook of 1855 he elaborated his opinions, denying the fixity of species and suggesting that smaller natural variations in populations might lead to sub-species and varieties but larger ones might lead to new species. He also summarised the work of Gaertner, who of course had different views. Although Mendel had left Vienna before the book was published, Unger's ideas must have been known to him (he in fact visited Vienna again in 1856). Mendel returned to Brno in 1853 and continued temporary teaching (having failed the exams again). However, in 1854 he bought seeds of 34 varieties of peas that he thought might show clearly marked differences in experiments. Soon he began to experiment with them. His motivation was most likely that he hoped to resolve the issue of the fixity of species raised in Vienna by Unger[32].

What little direct information there is on his motivation is cryptic: in his paper of 1865[33] he wrote that his experiments were relevant to:

> That, so far, no generally applicable law governing the formation and development of hybrids has been successfully formulated can hardly be wondered at by anyone who is acquainted with the extent of the task, and can appreciate the difficulties with which experiments of this class have to contend. A final decision can only be arrived at when we shall have before us the results of detailed experiments made on plants belonging to the most diverse orders.

> Those who survey the work done in this department will arrive at the conviction that among all the numerous experiments made, not

# GENESIS

one has been carried out to such an extent and in such a way as to make it possible to determine the number of different forms under which the offspring of the hybrids appear, or to arrange these forms with certainty according to their separate generations, or definitely to ascertain their statistical relations.

It requires indeed some courage to undertake a labor of such far-reaching extent; this appears, however, to be the only right way by which we can finally reach the solution of a question the importance of which cannot be overestimated in connection with the history of the evolution of organic forms.

Over the next fifteen years he continued teaching but made discoveries that were, eventually, to establish the foundation of genetics. He may have achieved even more but in 1868 he was elected Abbot and gave up experimenting (and teaching) soon after to concentrate on administration, particularly on fighting a new tax on monasteries. A serious smoker (he smoked 20 cigars a day), he suffered from poor health and died of chronic kidney disease on 6 June 1884. The composer Janáček, who after an earlier period as choirboy at the Abbey, was director of the organ school in Brno, played at his funeral.

His experiments on peas, *Pisum sativum* and its varieties and subspecies and closely-related species, were carefully planned and prepared. The choice of the pea as a subject was made because it was relatively easy to avoid self-pollination and hand pollination was straightforward if tiresome. But also Mendel wished to study the transmission of specific easily-recognised characteristics rather than the vaguer essences looked at by Naudin. The pea proved ideal for this, as Mendel confirmed in preliminary experiments, and he settled on seven distinct characteristics (which segregated independently) to study: seed shape and surface (smooth round or wrinkled and angular); ripe seed colour (yellow or green), seed coat colour (white or grey), pod shape (plump and smooth or skinny and wrinkled), unripe pod colour (green or bright yellow), location of flowers on the stem (along it or at the end) and stem length (2 m or less than 0.5 m). He kept 22 of the varieties as controls and then, from 1856, made a careful series of crosses with the remaining twelve to follow the seven characteristics he had identified. He made enough crosses to be able to draw statistically significant conclusions (some estimates suggest he grew nearly 30,000 plants in the course of the experiments).

The F1 hybrids were uniform showing the characteristic of one of the parents. So all the crosses between yellow seed parents and green seed parents gave plants producing yellow seeds, which meant that yellow seed

# GENES, FLIES, BOMBS...

was a *dominant* trait and green a *recessive* one. It was when the F1 crosses were self-fertilised that Mendel made his crucial discovery: he found that the recessive characteristic reappeared but was very nearly three times less frequent than the dominant one. So to every green pea plant there were three yellow ones: he counted 6033 yellow seeds for 2001 green ones. Similar results were obtained for all the characteristics and when he went on to self-fertilise successive generations (for 4-6 generations) he found consistent results.

## *MENDEL'S ANALYSIS*

He presented his results to two meetings of the Natural History Society of Brünn in early 1865 and they were published the following year in the Society's journal as *Versuche über Pflanzen-Hybriden*.

His explanation of all these results is in truth rather obscure – and historians have been arguing about his ideas for over one hundred years. He clearly thought that for each characteristic the pollen and egg cells could be of just two types which he called A(dominant) and a (recessive). The F1s between plants with different characteristics (Aa) would then always have the dominant trait. The explanation of the results of the F2 crosses is more difficult to follow. Some authors have seen in it something quite like the gene and its alleles as was later understood; others have seen rather less while still acknowledging Mendel's brilliant contribution. One point perhaps illustrates the difficulties. Mendel used just a single letter, A or a, to denote constant forms but Aa to denote the hybrid:

> The simplest case is afforded by the developmental series of each pair of differentiating characters. This series is represented by the expression $A+2Aa+a$, in which A and a signify the forms with constant differentiating characters, and Aa the hybrid form of both.

Supporters of the idea that Mendel actually had something like a gene in mind have argued that he simply abbreviated the AA and aa that would be used today to indicate the homozygote. Others have suggested that he had characteristics as the basic units of heredity. As we will see this is more a discussion about priority and Mendel's right to be called "the father of genetics" than about the science's development. What he certainly contributed was a brilliantly conceived and statistically sound[34] series of experiments that established the constant 3:1 ratio as something that had to be explained by everyone who followed him[35].

After the paper Mendel continued to work with his peas, beans and other plants. These included[36]. It later emerged that this was because of *Hieracium*'s ability to reproduce by parthenogenesis.

# GENESIS

After 1867 he stopped reporting what he had done and he ended his hybrid studies altogether in 1871 devoting himself after that to his abbot's work.

Some copies of his work were circulated as reprints (he seems to have asked for 40 but only four have turned up so far) and over the next 30 years or so it found its way into collections at over 50 learned societies and universities around the world. The Brünn journal was quite widely circulated for a publication of its time. His name was mentioned some 15 times in the literature as a hybridist[37].

However the significance of what he had found was not appreciated until 35 years after he presented his results in 1865 in what has become the most celebrated "rediscovery" of all science. The story of the isolated Silesian monk who had founded genetics but published in an obscure journal and was forgotten for several decade was born then. Perhaps Loren Eisley summed this view up best: "Mendel's only associates lived in the next century"[38]

# 2 EVOLUTION, CELLS AND CHROMOSOMES

It's just that sort of thing that makes a fellow chafe at our modern civilization and wonder if, after all, Man can be Nature's last word.

P G Wodehouse: *A Bit of Luck for Mabel*

### THE MULLERS TO 1900

In 1848, a few years after Mendel became a monk and a few before he bought his pea seeds, while Darwin struggled with his theory of natural selection, publication still a decade away, three brothers arrived in New York from Koblenz in Germany. Nicholas, Carl and Hermann Müller came to the USA for a new life, away from the constraints and inequities of the old Europe. Hermann went to search for gold in the California Gold Rush, which was in its early and hopeful days. He died there.

Nicholas and Carl stayed in New York, dropped the umlaut and opened the Muller Art Metal Works on Staten Island casting plaques, statuettes and elaborate clock cases under the name "C and N Muller". The creative side was initially down to Carl, a sculptor with a gold medal from the Paris World Fair to his name, but for some reason he quickly quit the business and America and returned to Europe. The company became "Nicholas Muller".

Over the following years Nicholas built up the business and produced, with his wife Johanna, four children: Johanna, Willy, Hermann and Otto. When Nicholas suffered a stroke and died in 1873, Hermann – who had studied law at City College and was now furthering his studies in Stuttgart – returned to take a hand in running the the business with Otto. The business name was changed to "Nicholas Muller's Sons".

The clocks and statuettes featured Cupids, dolphins, Roman soldiers, horses, hunters with dogs and many other designs and were noted for the quality and crispness of the casting.

Of the many clock designs produced between 1850 and 1900 – usually cast in an alloy of iron and lead – the most sought-after now is the Baseball Clock of about 1870 (known from just a few surviving examples) which has sold for many thousands of dollars.

Nicholas's descendents, apart from Willy's, distinguished themselves in various academic fields. Johanna mothered Alfred Kroeber after marrying Florence Kroeber and Alfred became a professor of anthropology at University of California. Otto's grandson, Herbert J Muller, became a professor of English at Indiana University. A distinguished historian and literary critic he is probably best known for his sweeping *The Uses of the Past*. Sadly Willy, who became a seaman, drowned off the coast of South America with, as far as is known, no issue.

Hermann, we should remember, was himself himself setting out on a career as an international lawyer before abandoning it to run the family business. He met and married Frances Lyons (whose family included a Canadian bishop and an English marquis and claimed some Jewish ancestry) and they produced first Ada and then, on 21 December 1890, Hermann Joseph Muller, future Nobel Prizewinner.

Hermann Sr quite possibly never fully settled to the business life. He certainly seems to have kept up a wide range of interests, political and scientific. Muller Jr recalled much later that his father:

> ... did much to imbue in me a strong sympathy for the working class, for oppressed peoples, and for internationalism, together with a scepticism of the righteousness of established governments and a hatred of imperialism (e. g. , against the U. S. aggression in the Philippines, British domination over the starving Hindus, etc.)

He also remembered them visiting the American Museum of Natural History when he was eight years old when his father:

> ... made clear to me, through the simple example of the succession of fossil horses' feet shown there, how organs and organisms became gradually changed through the interaction of accidental variation and natural selection (those which happened to have stronger middle toes, better adapted to escaping from the carnivores, tending to survive and so to leave more offspring, through numberless generations). And from that time the idea never left the back of my head, that if this could happen in nature, men should eventually be able to control the process, even in themselves, so as greatly to improve upon their natures.[1]

The Muller children were raised as Unitarians, a tolerant and accepting faith that emphasised the importance of reason and science, rejected notions of original sin and encouraged active involvement in politics. Unitarians have fought against slavery and for gender equality. The essential ecclesiastical requirement is a rejection of the Trinity but apart from that subscribers to all faiths are welcome. Muller declared himself a pantheist.

The company filed for bankruptcy in 1890, struggling to compete against lower quality products, and Hermann Snr died suddenly in 1900 but Frances continued her son's liberal home education as he attended local schools. Belts had to be tightened after the father's death but the young Hermann seems to have had a comfortable existence in a liberal environment.

## BEFORE DARWIN

As Mendel was conducting his experiments on peas, Charles Darwin was preparing his book *On the Origin of Species by means of Natural Selection, or the Preservation of Favoured Races in the Struggle for Life* for publication in 1859. Mendel read (or at least saw) Darwin's book before he made his presentation to the Brünn Natural History Society. Darwin just may have seen a reference to Mendel's work. If he had – and sought out the paper on peas and understood its significance – it might have provided a clue to solving one of the key problems facing natural selection. But we need to look back a while to understand that.

The Greeks were certainly concerned with generation but, the basis of so much of their thinking being geometry with its timeless and eternal truths, there was no real need to consider change at all. A living, harmonious and eternal whole left little scope for development; change would disturb the harmony. Aristotle, although a perceptive observer of nature, found fixed species that he expected to be eternal. Evolution had no place.

Christianity, as it spread, brought not a timeless cosmos but one with clearly defined limits. The cosmos had been created, it was reckoned from the genealogies listed in the Bible, around 4000 -6000BC[2] and it was to end on the day of judgement. As Mayr says "His making of the earth took six days, enough for all sorts of origins but not for any evolution[3]".

The simple picture began to break down as Copernicus, Galileo, Newton and all who followed showed that the cosmos was subject to physical laws. A Creator was required but, once his work was done, the universe spun on without further intervention. Men were released to look for more natural laws in more fields and this led them to seek non-supernatural explanations for more and more.

## GEOLOGY

The most significant science for evolution was probably geology. In the 18th century its students began to grasp how the earth had actually developed: extinct volcanoes were found and it was realised that the lava from them was the source of basalts, sedimentary rocks were identified and found to be enormously thick, suggesting a great age for the earth. The massive foldings of the sedimentary rocks that were found everywhere implied a violent past and the action of water in cutting valleys over long periods of time was quieter but hardly less dramatic. The earth began to seem very old indeed and Buffon calculated its age and published it in 1779 as at least 168,000 years (and privately thought it to be 500,000).

And then there were fossils. These had been known from Greek times and were explained away as shapes formed by the rock itself or as an outcome of spontaneous generation gone slightly awry. Christians explained them as the remains of plants and animals that died in the Flood; the biblical youth of the earth left little room for any other possibilities. This belief began to be challenged when different fossils were found to be present in different sedimentary rock strata (and were indeed a useful way of identifying them) and a sequence began to be built up. By 1800 William Smith, gathering information to draw his stratigraphic map of England and Wales (published in 1815)[4], had realised that the fossils associated with different strata were different. Georges Cuvier (1769-1832), in France, catalogued the fossils unearthed around Paris and his studies of the anatomy of living creatures allowed him to reconstruct the animals preserved as fossils. It was clear that some of the species that had been found no longer existed; Cuvier thought such disappearances the result of violent catastrophes rather than any gradual changes. Such discoveries posed a theological problem: either species had disappeared altogether or they had transformed into other species which still lived. The former would seem to violate the principle of plenitude; the latter would suggest permanent change in the natural world. How species originated or changed became, as we have seen, a major concern of the hybridisers.

While the idea that creatures could pass on to their descendents characters that they had acquired during their lifetimes was old, it has come to be associated with Jean-Baptiste Pierre Antoine de Monet, Chevalier de Lamarck (1744-1829).

## *LAMARCK*

Lamarck was the youngest of eleven children but all his brothers died leaving him with the family title but no money. Intended for the priesthood at 16, when his father died, he rebelled and ran away to the army. Commended and promoted for bravery in the Seven Years' War[5], he was forced to leave the service through illness when he was 22. Living on his pension and working as a bank clerk he spent much of his time

# GENES, FLIES, BOMBS...

studying botany and published a popular book on identification of flowers. This brought him to the attention of Buffon and he obtained posts in the botany department of the Jardin des Plantes in Paris. In spite of a lack of knowledge of zoology he was appointed to a chair in the subject in the new *Muséum national d'histoire naturelle* where he took an interest in invertebrates and became a self-taught authority. The story ended rather unhappily: he was married and widowed four times in his life, lost most of his children, became blind and was buried in a Paris pauper's grave.[6]

His celebrated (and notorious) contribution to biology was the promotion of the idea of the inheritance of acquired characteristics as an explanation of how organisms evolved – Lamarckism. It was based on the fact, obvious from geology, that the earth's environment had changed over time and this meant that organism had to change to adapt to them. Lamarck thought that organisms had a fundamental tendency to change and become more complex (it explained how lower forms of life – which originated in on-going spontaneous generation – became higher ones). However, in changing conditions, individuals adapted because organs they used more in the new conditions became bigger and stronger while those used less became smaller, weaker and withered. This was Lamarck's First Law. It implied no will; it was simply a result of physiological forces – which Lamarck detailed.

His Second Law said that these changes could be inherited. It was an old and widely-held view at the time and Lamarck must have felt no need to justify or explain it; he never did suggest a mechanism. The irony is that "Lamarckism" came to refer to just this part of his theory of evolution; it became synonymous with the inheritance of acquired characteristics (or "soft inheritance" as it became known). Lamarck's contribution as the first proponent of a plausible and scientific (if wrong) theory of evolution has largely gone unrecognised.

Lamarck's evolutionary ideas foundered on the rock of Georges Cuvier who was implacably opposed to evolution. It is possible that there was some religious reason for this [7]. Indeed he saw no evidence for much variation over time (other than as a result of catastrophes) at all: he stressed resemblance between parent and offspring and regarded any differences as superficial. He pointed to the animal mummies found in Egypt as showing no differences from the same animals found in his time. So his dismissal of Lamarck (and the other extant theory of Geoffroy) was based not on a rejection of soft inheritance but on opposition to evolution *per se*. However, Cuvier's standing was such that Lamarck's ideas were shunned until the second half of the 19th century.

Etienne Geoffroy Saint-Hilaire (1772-1844) was appointed to a professorship at the *Muséum* at the same time as Lamarck and in the 1820s expressed his own theory of how animals could be changed forever by changes in their environment. In this, called "Geoffroyism" by Mayr, there was a much more direct impact of the environment that through Lamarck's use/disuse mechanism. He thought that environmental changes caused changes in respiration which, in turn, required changes in the "respiratory fluids" and thereby affected the organism's structure. Creatures negatively affected would die out; those with beneficial adaptations would survive. Whether Geoffroy can be regarded as an evolutionist remains the subject of debate.

## *VESTIGES*

The controversies in France were hardly matched elsewhere except for what followed the publication of the anonymous *Vestiges of the Natural History of Creation* [8] *(Anon (Chambers R) 1844)*. There were many clear evolutionist ideas in the book based on the changing nature of fossils with time and the similarities between animals. The author also pointed out the progressively greater complexity of animals as time passed. The book contained a great deal of common sense but also many errors and no plausible mechanism causing the changes claimed was proposed. It was attacked by the biological establishment of the time but became a Victorian best-seller running to eleven editions by 1860. The author's identity remained a secret until 1871 when the he was revealed, posthumously, as Robert Chambers, the editor of *Chambers's Encylopaedia*. The book made no contribution to the evolutionary debate of the time but it was the first time the creationist interpretations of life's origins and developments were questioned in writing in England. It made a number of converts to evolutionist thinking including Alfred Wallace.

The evidence for evolution steadily increased. It started, ironically, with Cuvier's discoveries of diverging mammalian fauna in the Parisian fossils but there was much else. Geographic variation pointed that way as did the similarities found as comparative anatomy developed. The belief in catastrophes waned somewhat as creatures were found that had changed little in a very long time[9]. In spite of this many (if not most) of the leading authorities rejected the idea of continuing evolution, managing to square the emerging facts with their existing views, whether creationists with their stable earth or catastrophists. It needed a "cataclysmic event that would sweep the boards clean."[10] The publication of the *Origin of Species* in 1859 was that event.

## *DARWIN*

Charles Darwin (1809-1882) was born in Shrewsbury, Shropshire on 12

# GENES, FLIES, BOMBS...

February 1809. His father, Robert, was an eminent and wealthy physician, an FRS and son of Erasmus Darwin; his mother Susannah was the daughter of Josiah Wedgwood, the potter. Charles studied medicine at Edinburgh, as his father had done, but seems to have been horrified by the brutality of contemporary surgery. With a medical career unlikely he then went to Cambridge to study for the ministry; his aim was to be a country parson. He quite possibly saw this as a way to spend plenty of time on his consuming interest of natural history which he had developed at school and followed through university. His friendship with the Professor of Botany at Cambridge, the Reverend John Henslow, was the source of much of his education in natural history and he met other scientists who were to be important to him later. He graduated in 1831, spent some time on geology and was then invited, on Henslow's recommendation, as the naturalist on the survey ship *Beagle*. The ship left Plymouth in December 1831 and, after a cruise around South America and up to the Galapagos Islands, returned to Falmouth in October 1836 via New Zealand and Australia. On his return he studied the vast collection of specimens and notes he had accumulated as the ship surveyed the coast and islands of South America. The observations and reasoning that led him to his theory of evolution by natural selection have been analysed and speculated upon at great length in the literature, largely based on the notebooks he kept as he struggled with it. One fact that seems quite clear is that he had the key concepts in his mind by the end of 1838; Darwin himself describes a kind of epiphany he experienced on 23 September of that year after reading Malthus on population. It took more than 20 years to get from that point to the *Origin of Species*.

The essential elements of Darwin's theory were:

> Animals and plants have a much greater fertility than is ever realised. Fish produce millions of eggs; plants, vast numbers of seeds. Exponential explosions in populations are possible
>
> However, populations generally remain fairly stable
>
> Resources, mainly food, are limited
>
> There is therefore a struggle for existence
>
> Every population displays variability
>
> Some variations are more likely to survive and reproduce than others
>
> Much of the variation is heritable
>
> Natural selection will lead to change and new species

# EVOLUTION, CELLS AND CHROMOSOMES

He believed that the forces would act on individuals, gradually changing the population; there was no need for large sudden changes (saltations) or catastrophes. The theory implied a common descent for all creatures and plants and, of course, it had no place for the unchanging book of nature that had so constrained thinking. Darwin's theory of natural selection went on to become the dominant (some would say only) theory of evolution. There were modifications and the most important were associated with the sources of variation and the mechanism of heredity.

Darwin's view of the source of variation from the first edition of *Origin of Species* :

> I believe that the conditions of life, from their action on the reproductive system, are so far of the highest importance as causing variability. I do not believe that variability is an inherent and necessary contingency, under all circumstances, with all organic beings, as some authors have thought. The effects of variability are modified by various degrees of inheritance and of reversion. Variability is governed by many unknown laws, more especially by that of correlation of growth. Something may be attributed to the direct action of the conditions of life. Something must be attributed to use and disuse. The final result is thus rendered infinitely complex.

In Origins, Darwin clearly thought that acquired characteristics – some through use or disuse of organs – could be passed on to offspring. This view became stronger in the later editions as suggested by this extract from edition 6 of 1876:

> Changed habits produce an inherited effect, as in the period of the flowering of plants when transported from one climate to another. With animals the increased use or disuse of parts has had a more marked influence; .... The great and inherited development of the udders in cows and goats in countries where they are habitually milked, in comparison with these organs in other countries, is probably another instance of the effects of use. Not one of our domestic animals can be named which has not in some country drooping ears; and the view which has been suggested that the drooping is due to the disuse of the muscles of the ear, from the animals being seldom much alarmed, seems probable.

Darwin set out his theory of heredity in some detail in his 1868 book on variation under domestication[11]. The cells of different organs were thought to each generate miniscule "gemmules" which could either lead to the creation of new cells in the organ or spread around the body and then accumulate in the sexual organs. The gemmules could also be responsible for regeneration and parthenogenesis – as we have seen, major preoccupations of such theories. The sex organs, on this theory, would contain immense numbers of gemmules and each egg or sperm or pollen grain would hold an example of each gemmule. Sexual reproduction would then mingle those from the male and female, to form the basis for progeny.

This theory, which Darwin called pangenesis, suggested that the environment and the use to which organs were put would affect the gemmules that the cells in them produced. The accumulation in the sex organs would also be subject to the current conditions. Both these effects result in variation so, when the gemmules were passed on to the next generation, they would reflect the experience of the previous one (and preceding ones). Favourable variations would be passed on in a much greater degree than neutral or deleterious ones Hence a mechanism for natural selection. This was close to Lamarckism's inheritance of acquired characteristics but Darwin completely rejected Lamarck's ideas because he thought – wrongly – that the creature's will had a part in Lamarck's version. Summing up his own idea:

> On any ordinary view it is unintelligible how changed conditions, whether acting on the embryo, the young or adult animal, can cause inherited modifications. It is equally or even more unintelligible on any ordinary view, how the effects of the long-continued use or disuse of any part, or of changed habits of body or mind, can be inherited. A more perplexing problem can hardly be proposed; but on our view we have only to suppose that certain cells become at last not only functionally but structurally modified; and that these throw off similarly modified gemmules.[12]

The "provisional hypothesis" (Darwin's words) is advanced in detailed (if sometimes tentative) argument that addresses not just variation but many other matters such as regeneration, reversion and hybridisation. It was not entirely new – the Greeks had a version of it – but it was a material theory that did not rely on "the diffusion of mysterious essences and properties from either parent, or both, to the child."[13] It had a flaw in assuming transmission of acquired characteristics and another because it portrayed inheritance as a blending process. The fact that blending would tend to wipe out variations rather than allow them to be tested by natural selection was never really addressed. Of course, while Darwin may have got these detailed mechanisms wrong (to that extent he was a man of his time) his reliance on the fact of variation and inheritance from his studies of domesticated animals meant that his ideas were rather robust so he can be credited with one of the greatest intellectual achievement of the 19th century in his theory of evolution by natural selection.

## *WALLACE*

Something should be said about Alfred Russel Wallace (1823-1913).

The story is well-known. Darwin was slow to move from his initial ideas to publication as he accumulated more and more information supporting the idea of natural selection. In June 1858 he was still some way from being prepared to publish when he received a paper entitled *On the Tendency of Varieties to Depart Indefinitely From the Original Type* from Alfred Wallace who was working in Indonesia (the paper is referred to as the "Ternate Essay" after the island on which it was written), asking Darwin to pass it on to Charles Lyell. It was a bombshell because it succinctly outlined Wallace's own theory of evolution by natural selection – which was very similar to Darwin's and based on much the same evidence. Darwin was dismayed not least because Lyell had warned him of the risk of losing his priority after seeing an earlier paper by Wallace. It was agreed that there would be a joint presentation to the Linnaean Society of a paper by Darwin and Wallace's essay. This took place, with very little reaction on 1 July 1858.

Wallace did not try to explain the origins of variation, he just took it for granted and he had little to say about heredity – except for rejecting poor Lamarck.

The truly controversial idea of Darwin's was nothing to do with the mechanism of inheritance – he was after all close to the mainstream – but natural selection as a driving force. There were really three possible alternative evolutionary mechanisms to natural selection at the time.

> Lamarckism – now Neo-Lamarckism – was a widely-held alternative, seeing the inheritance of acquired characteristics as the dominant mechanism of evolution. The leading thinkers supporters were Ernst Haeckel (1834-1919), the German biologist, the British writer Samuel Butler and the US palaeontologist E D Cope. It faded as a respectable alternative by the end of the 19th century – but lingered on into the early years of the 20th.
>
> Religious alternatives saw a role for God in evolution. Some rejected natural selection altogether but others, such as the American botanist (and friend of Darwin) Asa Gray who believed that God had a role in creating the variations that natural selection worked upon. Such theistic interpretations dropped out of mainstream science by the end of the 19th century but retained

strong popular support from fundamentalist Christian religious groups into the 21st.

The third alternative to natural selection was orthogenesis, the idea that evolution progressed in an orderly, linear fashion driven by some external or internal force – although not necessarily to some plan and purpose. Prevalent through the 19th century, particularly among those who based their thinking on the fossil record, it waned rather by the end of the century as the very fossil record looked more complicated and full of dead ends.

There was an important modification of – rather than alternative to – Darwinism which was called saltationism. This proposed that new species were created – in one bound – by mutations. It was not based on a disagreement about heredity but was an important idea that was to carry us well into the 20th century. It is associated initially with Hugo de Vries but by the 20th century had notable supporters in William Bateson and T H Morgan.

The notion of the mutation is best left for a while as we follow the struggle between hard and soft inheritance after Darwin. It is easy to lose sight of the fact that the idea that acquired characteristics could be inherited – and the general soft inheritance principle that the genetic material was pliable and susceptible to environmental factors – was the dominant one through much of the 19th century. Mayr[14] has suggested that this really did not begin to change until the 1870s with the statement of the Swiss anatomist and embryologist Wilhelm His (1831-1904) in his 1874 book *Unsere Körperform und das physiologische Problem ihrer Entstehung*. Vogel, Leipzig.

> Until it has been refuted, I stand by the statement that characters cannot be inherited that were acquired during the lifetime of the individual.[15](translated from German)

His is remembered as the father of histogenesis – the science of the embryonic origins of tissues. He was the first to describe accurately the human embryo, was the inventor of the microtome (an invaluable device for obtaining thin slices of tissue for microscopy) and had a key role in identifying the remains of Johann Sebastian Bach[16].

## *WEISMANN*

It was August Weismann (1834-1914) who put forward most clearly the idea that Lamarckism had it wrong. He believed that inheritance in multicellular animals took place only through what he called the germ cells. These were a distinct line of cells having no interaction with the

## EVOLUTION, CELLS AND CHROMOSOMES

other (somatic) cells of the animal so were unaffected by them.

> Heredity depends upon the continuity of the molecular substance of the germ from generation to generation.

In his 1883 essay *On Heredity*[17] he rejected Lamarckian ideas and demolished the use/disuse mechanism and its role in Darwin's pangenesis.

> The inheritance of acquired characteristics has never been proved either by means of direct observation or by experiment. p80

His germ plasm theory clearly separated the somatic from the genetic (the germ) for the first time; the Weismann barrier, as it became known, prevented experience translating directly into inheritance. As a demonstration, he cut off the tails of 22 successive generations of mice and found no decrease in tail length in descendants as a result.

Although Weismann was so opposed to Darwin's idea of pangenes, he was a staunch supporter of evolution by natural selection at a time when this was in eclipse. His idea that natural selection accounted for evolution without Lamarckian mechanisms (which was shortly called Neo-Darwinism) can be seen as the source of the version of Darwinism that remains the dominant theory of evolution today. Mayr regards Weismann as second in importance as an evolutionary thinker only to Darwin himself.

There was much about the detail of Weismann's theory that proved to be incorrect but the blow it dealt to the idea of acquired characteristics was fatal – although it proved slow to actually die.[18]

### *THE CELL*

Ideas about the cell were having an effect on thinking about heredity by the 1870s. They reached back to Hooke and his *Micrographia* of 1655 where he had described ("like a Honeycomb") and drawn the cells of cork and of the leaf of a stinging nettle. Others saw similar structures but progress was inhibited by the poor quality of the microscopes available. Improved instruments, corrected for both spherical and chromatic aberrations, began to become available in the 1830s and this created a boom in microscopic studies in biology. Cells had been studied before that but much time was wasted because of the poor instruments and people chased what turned out to be diffraction haloes and other purely optical phenomena. That is not to say there were no discoveries of worth; Robert Brown (of the eponymous motion) saw the nucleus for the first time in 1831 in the cells of orchids. Jan Purkyně and his group at the University of Breslau (now Wrocław in Poland) had seen cells and nuclei

# GENES, FLIES, BOMBS...

in both animals and plants and referred to the similarities between them, supporting the general notion of a cell.[19]

However, the two people usually credited with formulating the cell theory are Matthias Schleiden (1804-1881) and Theodor Schwann (1810-1882). Schleiden proposed in 1838 that all plants were composed of cells; Schwann extended this idea to animals in the following year. Their model had a cell with walls, full of protoplasm with a nucleus. It was not understood how cells reproduced. Schleiden had introduced a substance he called the cytoblastema. Itself formless and structureless, it was the source of new nuclei and hence new cells, which arose within the cell by a process like crystallisation. Schwann thought the new cells arose from this potent fluid outside existing cells. Both views implied a kind of spontaneous generation – although Schleiden and Schwann denied this. The cell theory, as a basic description of the building blocks of plants and animals, was widely accepted as the basis for further research so the misunderstanding of cell genesis was quite quickly corrected. By the 1850s Rudolf Virchow (1821-1902), Robert Remak (1815-1865) and Albert Kölliker (1817-1905) described mitosis. It seems that Remak was first but that Virchow was a better publicist[20]. Virchow coined the phrase *omnis cellula e cellula* (all cells come from existing cells) and, if he did not actually have priority in discovering mitosis, he was the person who did most to establish a coherent cell theory through his writings in the late 1840s and the 1850s. He laid the foundations for modern cellular pathology, for the first time seeing that diseases, including cancers, should be treated as the result of cell malfunctions. His contribution was not restricted to theory: he set up hospitals and hospital trains during the Austro-Prussian War of 1866 and the Franco-Prussian War of 1870-1. A distinctly liberal man, he was a member of Berlin City Council for over 50 years where he promoted social and public health improvements. He served in the Prussian parliament as a leading progressive and a vigorous opponent of Bismarck (whom he is alleged to have caused to back down from a proposed duel) and, from 1880 to 1893, in the Reichstag.

It became apparent that protoplasm (which was soon, following Kölliker, called "cytoplasm") was not a homogeneous clear jelly but had structure within it. This could not be seen well at the time, even though microscopes continued to improve, because the very methods of staining used to visualise cell structure produced confusing artefacts. So it was the 1870s before the oil immersion microscope and better staining techniques showed microscopists the organelles of the cell. Mitochondria were seen first in 1898 as was the Golgi apparatus[21].

# EVOLUTION, CELLS AND CHROMOSOMES

*THE NUCLEUS*

The nucleus began to show internal structure and the objects that appeared during mitosis (which had been seen in the 1840s) were studied more closely. Several authors described them segregating between the daughter cells. Eduard Strasburger(1844-1912) published a milestone book on cell formation and division in 1875 which recognised the similarity in mitosis in plants and animals – an important unifying observation. The Belgian Edouard van Beneden (1846-1910), for example, in 1883 observed the various stages of mitosis in *Ascaris megacephalus*, a parasitic worm of the horse that conveniently has large transparent eggs that shuffle down the uterus after fertilization offering a readily-seen sequence of development. Walther Flemming (1843-1905) stained the chromosomes of a salamander (he called them "chromatin" because they stained well) in the late 1870s[22] and, in 1880, made the crucial observation that they split longitudinally in mitosis and separated; precisely sharing each chromosome between the daughters. He coined his own phrase: *Omnis nucleus e nucleo*.

It is quite remarkable that most of these discoveries were made by German biologists. Schwann, du Bois-Reymond, Remak, Kolliker and von Helmholtz (as well as other important figures we have not mentioned) were all students of Johannes Peter Muller at the University of Berlin so Muller must rate as the grandfather of cellular biology. Matthias Schleiden worked in Jena, Strasburger although Polish by birth worked in Bonn, Virchow in Berlin and Flemming was at the University of Kiel. August Weismann can be added to the list. The hegemony of Germany in the biological science and medicine was to last until the First World War.

The role of the chromosomes in heredity emerged slowly. Ernest Haeckel had suggested in 1866 that the nucleus was responsible for the transmission of characteristics but the idea was so buried in other erroneous musings that it was never taken up. Karl Wilhelm von Nägeli (1817-1891) put forward a theory in 1884 that the information was carried by the "idioplasm", a component of the cytoplasm. This came in equal parts from the mother and father, was made of strands each with specific properties and controlled the differentiation of tissues. It sounds quite a modern idea but would be better characterised as an imaginative one with slight observational basis. Mayr[23] is to the point: "...almost every detail of his theory was radically wrong and almost none of it based on any known fact." Nägeli himself was rather pleased with it: "The idioplasm theory...permits the only possible interpretation how inheritance and phylogenetic change can take place naturally, that is mechanically." Because of his status the theory had widespread if short-lived influence. It had no role for the nucleus. It was, of course, Nägeli who had allegedly diverted Mendel from further productive work on peas onto less tractable subjects.

Wilhelm Roux (1850-1924) gave, in 1883, the first good reason for thinking that heredity might be strongly linked to the chromosomes when he saw that the longitudinal splitting of them resulted in identical copies being shared between daughter cells. Each chromosome could thus carry genetic information which could be accurately shared between the daughter cells.

The role of the nucleus was also clearly shown in the experiments with orchids conducted by Strasburger in 1884. He showed that only the nucleus of the pollen is forced into the embryo sac. So only the nucleus was involved in further development and was thus the sole source of hereditary determinants. Oscar Hertwig (1849-1922) in 1884 and 1885 thought that the nucleus might be involved and speculated that the "nuclein" extracted in 1869 by Miescher might be the carrier of genetic information.

## BOVERI

Theodor Boveri (1862-1915) made important observations from 1887 to well into the twentieth century. He showed that the number and structure of chromosomes were preserved in *Ascaris* through cell division: the chromosomes of the daughter cells looked just like those of the parent. So chromosomes were independent entities that were preserved even when they disappeared from view in resting cells. Roux had suggested that individual chromosomes had specific genetic roles – so all were needed for proper development – but this was largely speculation (if inspired such) and not decisively supported by observation or experiment.

In 1889 Boveri turned to sea urchin eggs as a study material: Hertwig had used them a decade earlier for his fertilisation studies and they were transparent, readily available and easy to work with. Hertwig had shown that if he fractured the eggs of sea urchins by shaking them, fragments without nuclei could be fertilised and would begin to develop. Boveri now extended this by fertilising the eggs and the nucleus-free fragments with sperm from a different species with distinct-looking larvae. He found that while there were hybrids (as he had found earlier with untreated eggs) some of the resulting larvae looked very like the paternal larvae. So it seemed that a single (haploid) set of chromosomes from the father was enough for development but still the role of the individual chromosomes was not clear. It later emerged that the shaking did rather more damage than thought at the time and the results of the imaginative experiment were questioned.

His decisive experiments were conducted between 1901 and 1907 with a first report in 1902. It was known that the fertilised egg of the sea urchin would sometimes split into four rather than two at its first cleavage. Such eggs would not develop into larvae. Boveri had seen similar behaviour in *Ascaris* and identified the reason: these were eggs fertilized by two sperms. Since each sperm introduced two centrosomes, the anchor points for the spindle that pulled the chromosomes apart in mitosis, the abnormal eggs would have four centrosomes. Since there would be three sets of chromosomes in the egg, two paternal and one maternal, and they would be pulled four ways the likelihood was that none of the four daughter cells would have a full complement. Very few of them would be viable, as he found. This was interesting in itself but Boveri went a step further. He found that he could produce eggs that split into three rather than four if they were shaken after the double fertilization: this stopped one of the centromeres dividing and there were thus only three in the cell. Following the development of these cells he found that they gave a much higher proportion of relatively normal embryos than those that had split into four.

He found that he could explain these results if chromosome pairs were split randomly between centromeres. Having measured the frequency of the normal embryos from the three and four cell splits he was able to explain them quite accurately with a simple probabilistic model of chromosome partition between centromeres – a model realised with numbered wooden beads to represent chromosomes and trays to represent daughter cells. It could only mean that individual chromosomes had different roles in development as Roux had suggested nearly 20 years before: at least one of each type was essential. [24]

Boveri continued his work with his wife Marcella but died relatively young after becoming infested with *Ascaris* worms. As he wrote just before he died: "It is mean when the beasts you have worked on, now start working on you."[25]

Of course while Boveri made his observations and did his experiments other workers were not idle. T H Montgomery in 1901 drew a strong inference that individual chromosomes differed in their qualities – but it was based on much speculation. A more solid indication came from C E McClung's work at the University of Kansas with grasshoppers. He found that only half of their sperm had a small accessory chromosome. He deduced that it was involved in determining gender of offspring: a very specific function for at least one of the chromosomes. But it was McClung's student Walter S Sutton, working as a graduate student with E

# GENES, FLIES, BOMBS...

B Wilson at Columbia University in 1902 and 1903, who extended the work that had been done and produced a coherent theory of chromosomes and their functions. He saw in grasshoppers (*Brachystola magna*) the identical pairs of chromosomes retaining their distinctive appearances through several generations of division and saw their splitting in meiosis. He summarised his views in his 1902 paper[26].

> We have already reviewed the reasons for believing the accessory chromosome in the cells of *Brachystola* to be the possessor of specific functions and it only remains again to call attention to the likelihood that the constant morphological differences between the ordinary chromosomes are the visible expression of physiological or qualitative differences.

The final paragraph was equally significant:

> I may finally call attention to the probability that the association of paternal and maternal chromosomes in pairs and their subsequent separation during the reducing division as indicated above may constitute the physical basis of the Mendelian law of heredity. To this subject I hope soon to return in another place.

## SUTTON-BOVERI THEORY

He did return to this in his 1903 paper[27] where he expounded the chromosome theory with great clarity and demonstrated that his closing remarks were justified. The theory could fully account for Mendel's Laws with what he had seen of chromosome duplication and segregation and on the assumption that each chromosome carried determinants for many characters. He explained the phenomenon of gene linkage in which different characters were sometimes inherited together and suggested that what seemed like non-Mendelian blending inheritance could be the result of several genes working together. He knew, and fully acknowledged throughout, the work of Boveri on chromosome continuity and individuality.[28]

The theory he put forward became known as the Sutton-Boveri hypothesis through the influence of Wilson, who had links with Boveri: Miss Marcella O'Grady, one of his students, had moved to Wurzburg to work with Boveri and subsequently married him. Walter Sutton gave up his genetics researches altogether soon after the second paper was written He worked as a foreman in the US oil fields where he showed himself a talented engineer and inventor but when he had accumulated enough money he trained to be a surgeon where he proved to be capable and inventive. He died in 1916 at the age of 39 of a ruptured appendix. The Sutton-Boveri theory, in spite of some opposition from prominent people

# EVOLUTION, CELLS AND CHROMOSOMES

such as T H Morgan, stood the test of time.

The link between chromosomes and Mendel was a key step in the heredity story. The laborious and innovative work of cytologists slowly revealed the secrets of the nucleus but Mendelism advanced in a flash after decades of obscurity in remarkable rediscovery in 1900.

## *MENDEL REDISCOVERED*

One of the unsolved problems of Darwinism in the 1890s was posed by the blending theory of heredity that was commonly held. If evolution were to take place as Darwin had suggested by the action of natural selection on small variations then how did they survive being swamped by all the normality that was around? Even if some small variation survived one generation why would it not then quickly be wiped out? One of the people who addressed the problem was Hugo de Vries, who had become convinced that the dominant mechanism was not selection for small changes but for big ones. He had seen sports or mutations particularly in the Evening Primrose and these seemed to survive through several generations. So perhaps evolution progressed by leaps rather than small steps. The version of Darwinism developed from this idea became known as saltationism and one of the most influential books was de Vries's *Die Mutationtheorie* of 1901.

De Vries published his *Intracellular Pangenesis* in 1889, expressing his view that characters were inherited independently, and in the early 1890s he set out on an experimental programme to demonstrate this and investigate variation and its inheritance. He found segregation in more than 30 plant species and in F2 crossing experiments on more than a dozen different plant species found ratios close to the 3:1 ratio that Mendel had found. He published his results in spring 1900 in three papers. Although from later accounts he was clearly aware of Mendel's work at the time, he did not mention Mendel at all in the first, in the second only briefly and in the third rather fully. Some have interpreted this to mean that de Vries hoped to get away without mentioning Mendel and taking all the credit but was thwarted when he realised that Correns and Tschermak had let the cat out of the bag.

When he received a reprint of de Vries' first paper Carl Correns (1864-1933) was surprised. He had submitted a preliminary account of his work (in which he referred to Mendel) on the colour of hybrid peas to a journal in December 1899 and it was published in January 1900. He had become aware of Mendel's work in late 1899 after obtaining similar results in experiments on pea hybridisation. When he saw the reprint he quickly wrote a short summary and submitted it to a journal where it was published in May 1900. It fully credited Mendel with a title translated as: "Mendel's laws concerning the progeny of varietal hybrids." It is intriguing that Correns had been encouraged to take up botany by Karl Nägeli, the man who corresponded with Mendel and turned him away from further work on the pea to other less suitable species. Nägeli was a family friend and Correns married his niece. No evidence has come to light suggesting that Correns knew of Mendel through Nägeli. Indeed, if he knew about Mendel's work he was very slow in taking advantage of the fact.

Erich Tschermak-Seysenegg(1871-1962) also had a link with Mendel: his grandfather Eduard Fenzl had taught him in Vienna. Tschermak's interests were in practical plant breeding, improving stocks through hybridisation and inbreeding. He too had been crossing peas and found the same ratios as Mendel. He wrote two papers in 1900. The first of these showed less than full understanding of Mendel's principles and the second, while better, was written after reading de Vries' and Correns' papers. Although Tschermak is usually credited as a co-discoverer, historians continue to debate whether this is too generous.

## BATESON

William Bateson (1861-1926) had been conducting breeding experiments through the 1890s and found many results we can now see showed Mendelian inheritance. However he did not actually formulate a theory of inheritance so reading Mendel's original paper in May 1900 was a revelation. Prompted by a reprint from de Vries, he first saw it on a train from Cambridge on his way to deliver a paper in London and promptly changed his talk to the Royal Horticultural Society to cover it.

He quickly became a most enthusiastic promoter of Mendel (much more so than de Vries, Correns and Tschermak) and arranged for the paper to be translated and published. This was partly because he thought that Mendel's ideas gave support to the notion of saltationism as a mechanism for speciation, which he shared with de Vries. A vigorous controversialist when needed, he invented the terms "genetics", "allele", "homozygote" and "heterozygote", was the major force in the development of the new science and was called "the real founder of the science of genetics" by Castle.[29] His 1902 book *Mendel's Principles of Heredity*, which had translations of the original papers, was a vigorous promotion and defence that introduced many to Mendel's ideas. His influence waned rather after about 1910 because of his continued opposition to the chromosome theory of inheritance, which was becoming widely accepted. However, he continued with his insistent promotion of saltationism in the debate with the gradualists that was to go on for many years after his death.

# EVOLUTION, CELLS AND CHROMOSOMES

## *PRESENCE-ABSENCE THEORY*

Bateson is inextricably linked with the presence-absence theory of genes. This proposed that an interpretation of Mendel's results was that genes were either present or absent from the germline. Dominance was an expression of presence; recessiveness of absence. Mutations were, being generally recessive, removals of genes. It was not original to Bateson – Correns and Cuénot both suggested versions of it – but he was its loudest proponent from about 1904. The theory proved surprisingly resilient and survived until the 1930s. One of the major problems it faced was that it suggested that evolution proceeded by the loss of genes, which seemed unlikely, but many other objections to it were accommodated by adjustments. It was finally dealt mortal blows by the observation of reverse mutations and multiple alleles[30]. Even in 1935, in the landmark paper that signalled the beginning of molecular biology, Timoféeff-Ressovsky felt it necessary to refute it.[31]

## *WELDON AND GALTON*

However, there were opponents of Mendelism. W F R Weldon (1860-1906), a professor at University College, London was a conventional Darwinian and believed that evolution occurred through continuous variations and this seemed incompatible with Mendel's ideas. He conducted a bitter feud with Bateson until his death in 1906.

Francis Galton (1822-1911) was Charles Darwin's half-cousin and like Darwin gave up studying for a medical career because he found it unpleasant. Four years studying mathematics at Cambridge resulted in mental breakdown and a mediocre degree but his private income allowed him to indulge interests in meteorology and statistics as well as heredity. Although he had a hand in setting up the UK Meteorological Office and made major technical advances in statistics, he is now remembered best for his application of statistical methods to the study of heredity through the study of measurable traits. In his book *Hereditary Genius* of 1869 and elsewhere he presented a version of blending inheritance which is now called "Galton's Ancestral Law of Inheritance". In this it was claimed that our genetic inheritance came ¼ from our parents, 1/8 from grandparents, 1/16 from great-grandparents and so on. Nothing was completely lost and even remote ancestors might have a dramatic effect from their small contribution. His views of the mechanics of inheritance were somewhat different from Darwin's ; he thought the hereditary material was carried in the reproductive tissues with little or no contamination from the other parts of the body. He performed blood transfusion experiments with rabbits to show that there were none of Darwin's "gemmules" circulating around the body[32]. His Ancestral Law had been derived on the basis of measurements of continuously varying quantities such as height, intelligence and strength and it seemed quite at variance with Mendelism. In evolutionary terms he was however a saltationist: his studies led him to believe small changes would be blended away with no chance of speciation.

The obsession with measurements led to Bateson, Weldon and others being called biometricians. Weldon and Carl Pearson, with support from Galton, founded the journal *Biometrika* in 1901 and it became a vehicle for attacks on Mendelism. This, with its obsession with discrete rather than continuous characters, had to be fought.

The evidence for Mendelism in a wide variety of plants and in animals grew and supposed examples of non-Mendelian inheritance were explained using the theory. The inheritance of cocks' comb structures is one notable example from the time.

### *CONTINUOUS VARIATION*

Even as early as 1902 there were hints of how the differences between the Mendelians and the biometricians might be resolved: the statistician George Udny Yule (1871-1951) pointed out the possibility that factors that seemed non-Mendelian, continuous and the result of blending inheritance could in fact be controlled by many genes (as we would now say) working together.[33]

# 3 FLIES, GENES AND MUTATIONS

*MULLER 1900-1915*

The Mullers moved to the Bronx when Hermann was two years old and then, in about 1896, to the Upper East Side to an apartment on 122nd Street near 7$^{th}$ Avenue. He attended elementary schools (PS10 and PS89). In 1903 he went to Morris High School, the first High School in the Bronx. For the next four years he walked across the bridges over the Harlem River, up Boston Road, to the architectural glory that was (and until recently still was) the School's home. It was here that he met Edgar Altenburg.

Altenburg was the son of a piano maker and shared something of Muller's background. Frederick Ernst Altenburg had founded the business in Germany in 1847. His son Frederick K moved, with his family, to the USA in 1855, just a few years after the Muller brothers. They settled in New York and Frederick and his sons set up the family business, with a small factory in the Bowery and soon two shops in the city. In 1908 or thereabouts the business moved to New Jersey and by then it was being run by Edgar's father Gustave and his uncle Otto. Edgar's. His brothers Alexander, Otto and Herbert all worked for it. The family became well-known for designing and building the upright pianos[1] that replaced the earlier square type. The business survives today in the same premises in New Jersey, but now selling Altenburg pianos made in Korea and China.

As well as sharing backgrounds Edgar and Hermann shared a passion for science and, it seems, even at this young age, a political and philosophical outlook. Edgar persuaded Hermann to become an atheist and strengthened the liberal or radical views Muller had imbibed from his mother and father. Between them they founded the Morris Science Club: Muller gave talks on flying and the planet Mars, based largely on his extensive collection of clippings about scientific matters and world events. For his graduation exercise in 1907 he chose "The Need for Higher Ideals in Business and Politics."

He arrived at Columbia University in the autumn of 1907. Younger than the average freshman at 16 and much shorter than most at 5 ft 2 in (he was to grow no more), he had won a scholarship (the Cooper-Hewitt) that took him to the Liberal Arts school. He was joined after a year by Edgar Altenburg. In his freshman year he studied general biology with J H MacGregor and MacGregor neither failed to teach with a strong evolutionist flavour nor to make it plain that Columbia had a leading role in the new experimental approach to biology. It must have inspired Hermann. It was over the first summer, when he worked as a hotel clerk in New Jersey and read voraciously at quiet times, that he developed his passion for biology. When he returned to school in the autumn he kept reading and took courses under E B Wilson, departmental chairman and a leading proponent of the role of the cell nucleus in development and inheritance. One of Wilson's students, Walter S Sutton, had recently demonstrated the link between chromosomes and Mendel's new recovered laws. Muller came into contact with a relative newcomer to the Columbia faculty, T H Morgan. Morgan had come to Columbia from Bryn Mawr in 1904 as Professor of Experimental Zoology to continue his studies on pigeons, rats and mice intended to throw some light on the creation of species.

In the autumn of 1909 Muller was so committed that he founded a Biology Club. The members included not only Altenburg but two students from the year below him: Calvin Bridges and Alfred Sturtevant. He gave a lecture to the Club: *Basic Principles of Life Processes*.

He must have made some impact because he was invited to join the Peithologian Society (*Vitam Impendere Vero* – To Devote One's Life To Truth) which had been founded in 1806. He and Altenburg joined the Intercollegiate Socialist Society.

He graduated *cum laude* in 1910 (having squeezed four years work into three) and was elected to the prestigious Phi Beta Kappa Society.

There were no immediate graduate opportunities for him in zoology at Columbia so he spent a year on a scholarship in the Columbia Medical College working on the physiology of nerve impulse transmission and obtaining a master's degree for it. Another year was spent at Cornell Medical College teaching some classes and working on creatinine metabolism – but that came to nothing. Throughout he was teaching English to foreign students to get some money. It was a punishing routine which began to affect his health. However he had kept his contacts with T H Morgan and his group through attendance at Morgan's lectures and through Bridges and Sturtevant. He kept in contact too with Altenburg. His opportunity came in 1912 when he obtained a teaching post at Columbia and a chance for work with Morgan. Morgan had, by then given up pigeons, rats and mice and was working with Bridges and Sturtevant on a tiny fly.

# FLIES, GENES AND MUTATIONS

## *MORGAN'S SLOW EPIPHANY*

In the USA Charles Otis Whitman (1842-1910), the first Director of the Woods Hole Laboratory and just possibly the USA's most influential biologist of the time, was vigorously opposed to the Mendelian idea largely because he had studied heredity in pigeons and doves (he is supposed to have studied over 700 species). As Sturtevant said "...in pigeons you get, to put it mildly, a mess". There were no simple ratios here and Whitman became an adherent of orthogenesis.[2]

Morgan, a regular worker at Woods Hole, also found Mendel difficult to swallow – along with Weismann's ideas of heredity, Darwinism, Lamarckism (he preferred the de Vries mutations explanation) and the chromosome theory.[3] All lacked what he saw as a convincing experimental base. He regarded the suggestion that multiple genes were responsible for characters and might explain results like those of Whitman as special pleading. The chromosomes as carriers of genetic information seemed to him to imply linkages of characters in inheritance not actually seen. He set out his objections in some detail in a paper in 1910.[4] However by 1915 he had, with colleagues, written a book[5] that was to be the foundation of modern ideas of genetics and was awarded the Nobel Prize in 1933 for his work linking chromosomes and heredity.[6] This dramatic turn-around was down to a small fly.

## *DROSOPHILA*

*Drosophila melanogaster* is a minute fly with a taste for the yeasts growing on over-ripe fruit; its name, which means the black-bellied dew-lover, is something of a misnomer About 3 mm long, it is yellow-brown in colour with black rings across its abdomen and, normally, terracotta-coloured eyes. The male, which is slightly smaller than the female, develops a black patch on its abdomen as it matures and the two sexes are quite easily distinguished at birth: the male has sex combs (bristles on its front legs) and claspers (more bristles) which allow it to clamp itself to the female while it mates. It breeds very quickly. From egg to adult via lava and pupa takes just over a week under the right conditions and the young females are receptive within a few hours of birth. After mating the female produces around 100 eggs a day for the remainder of her 30 day lifespan so the generation time can be 10 days or less. The female generally mates with several males and the sperm is stored. The last male mating is most likely to father young either because of segregation of sperm in the female or because males somehow manage to neutralise the sperm of their predecessors. The sperm of *D melanogaster* is remarkable in being around 1.8 mm long when uncoiled – something over half the fly's body length.[7]

# GENES, FLIES, BOMBS...

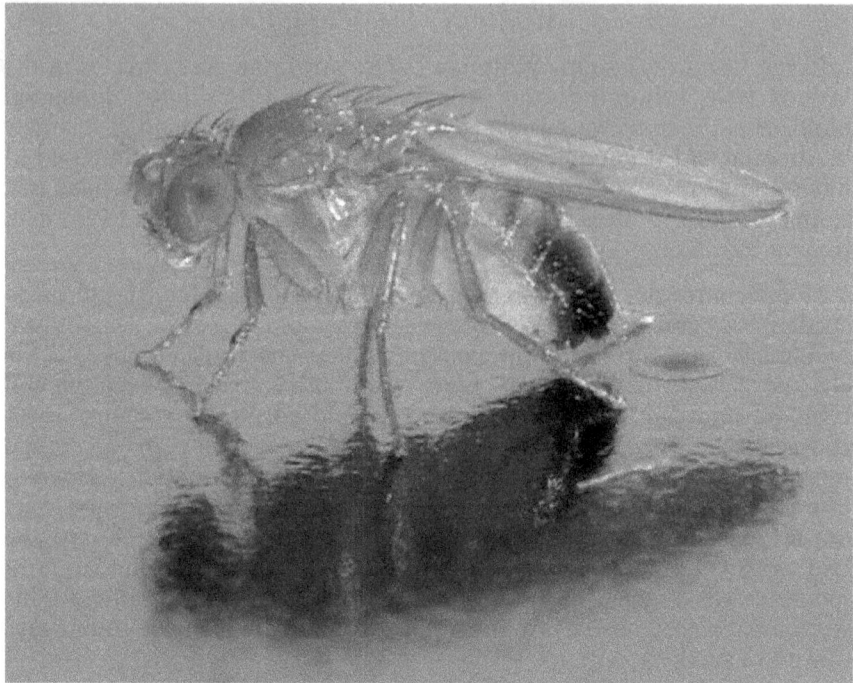

Illustration 4: *Drosophila melanogaster* (A Karwath)

*D melanogaster* is just one of around 1500 species of fruit fly catalogued (there may be many time more) in the genus which is very varied and found in many habitats around the world where they can feed on decaying fruit and fungi. It is often found associated with man and probably entered the USA in fruit shipments from tropical areas.

It seems to have first been suggested as a laboratory animal to W E Castle around 1901 by the entomologist C W Woodworth, who bred it in quantity at Harvard. Castle, an embryologist who had turned to genetics and had been working there with guinea pigs, began his studies of inbreeding with the insect and published his results in 1906, meanwhile F W Carpenter used the fly in behavioural studies at Harvard and published in 1905 – the first account of the use of the fly. Fernandus Payne, working for Morgan at Columbia, started experiments in late 1907. Like most others he seems to have collected the flies himself: L S Quackenbush, also at Columbia, used insects collected at Wood Hole.[8]

The little fly proved an ideal subject for breeding experiments: the generation time was short, it was easy to rear; the females were fecund

# FLIES, GENES AND MUTATIONS

and could easily be identified and segregated at birth for controlled mating. Morgan, stimulated by de Vries' mutation experiments with Evening Primrose, had done some experiments with mice and rats but soon realised the advantages of *Drosophila* and started his own breeding trials in 1909 – with, it seems, the aim right from the start of studying mutations.

## THE FLY ROOM

Morgan's laboratory was a small room, "The Fly Room", at the top of Schermerhorn Hall at Columbia University on the Upper West Side of Manhattan. Room 613 was just 5m x 7m and contained eight desks for students and visitors: Morgan had a small office next door.

The room was lined with half-pint milk and cream bottles – some purloined by students – with a paste of rotting bananas at the bottom (a rather more refined food was used later). In each bottle were flies trapped by a cotton wool plug. The room had a characteristic smell of fermenting bananas. The flies were examined with a hand lens or jeweller's loupe in the early days but stereoscopic microscopes were introduced later on. Individuals were isolated from their siblings by taking an empty bottle and allowing flies to crawl into it from the original bottle until there was an equal number in each. The bottles were scanned with the lens to find the wanted fly and the whole process was repeated. Eventually the required creature would be separated out.

Two key early occupants of the desks in the room were Sturtevant and Bridges.

## MORGAN'S MUTATIONS

Morgan tried, with Payne, several methods for inducing mutations in his subjects: chemicals and X-rays and radium radiation. He had no success but in 1910 did find (or some say his wife found) a white-eyed mutation in a single male fly. When he crossed this with wild (red-eyed) females he obtained no white-eyed females and very few white-eyed males in the first generation. When he went on to cross members of the first generation he found 2500 red-eyed females, no white eyed-females, 1000 red-eyed males and 782 white-eyed males in the next generation. This was clearly at odds with simple Mendelian scheme and was sex linked.

Morgan went on to find more mutations, notably what became known as "rudimentary" – referring to the fly's wings – which was also sex linked. These two mutants formed the basis for an important test of earlier ideas about the chromosomes.

One of the problems that had been appreciated for some time regarding the chromosomes as the carriers of the genes was that there seemed to be more genes than chromosomes. If this were so then many genetic traits should be transmitted together – linked. There was already some evidence that this was the case.

Linkage had been found by Correns in 1900 in stocks. Incomplete linkage in the sweet pea had been seen by Bateson and Punnett in 1905 but they were led down the wrong path by some spuriously regular-looking results and rejected the chromosomal interpretation. They favoured what they called "coupling" and "repulsion" – and later, in a variant form, "reduplication". The first reliable examples of linkage came when traits linked to the sex of the subject were found.

### LINKAGE - MOTHS AND CANARIES

The currant moth *Abraxus* normally had quite strong colours but a variety *A lacticolor* exists with a rather washed-out appearance. Doncaster and Raynor in 1906 found that the inheritance pattern of the variety is sex-dependent: reciprocal crosses (ones with *lacticolor* male x normal female and *lacticolor* female x normal male) give different results and most *lacticolor* are females. In 1908 two women, Florence Durham (Bateson's sister-in-law) and Dorothea Marryat, studied the inheritance patterns of green and cinnamon canaries. The green canary is the wild type and the green is the result of black and brown feather patterns overlying a yellow background. The bright yellow we associate with canaries comes from breeding out the overlay colour. The cinnamon variety is a rather dull variety which has just brown overlay (breeders intensify the colour by feeding). The green canary has black eyes while the cinnamon has pink, unpigmented ones when it is young. Durham and Marryat showed that the pink-eyed cinnamon variety was inherited in a sex-linked way and reciprocal crosses gave different results.

### CROSSING OVER

These were important examples because they supported the chromosome notion but hardly helped with the question of chromosome and gene numbers. Of course it was also significant that linkage was not seen more widely – if there were many genes on the same chromosome then they should all be linked. The problems were first recognised by Sutton and de Vries in 1903 and de Vries put forward a possible solution: genes could be swapped between homologous chromosomes in meiosis. Boveri came to a similar conclusion the following year.[9]

### CHROMOSOME MAPS

It was Morgan's experiment with his two mutant flies that began to clear up the confusion. In crossings he found results that were explicable only if there was an exchange of genetic material (recombination) between the sex chromosomes of the female. He could tie this to the cytological work of Frans (F A) Janssens (1863-1924) published in 1909. Janssens, a biologist and talented microscopist at the University of Leuven in Belgium, had seen chiasmata (crosses) where he thought homologous chromosomes which had replicated in meiosis might exchange fragments of material. Morgan and Eleth Cattell later[10] called this crossing over.

This work opened up a further possible line of research that would surely make the chromosome the undisputed carrier of genetic information. Morgan soon realised that, if genes were strung out along a chromosome, the chance of the linkage between two of them being broken by crossing over depended upon how far apart they were. The linkage between genes close together would be much higher than those far apart. In 1911, after a discussion with Morgan, Alfred Sturtevant, then an undergraduate, had the insight that he could use the linkage data they had accumulated for the five sex-linked genes of *Drosophila* they had found so far to construct a map showing the locations of the genes on the chromosome. He did just this overnight in late 1911 and published the first results for six genes in 1913, taking account of the possibility of double crossing over in the process. The wider implications were clearly recognised:

> These results are explained on the basis of Morgan's application of Janssens' chiasmatype hypothesis to associative inheritance. They form a new argument in favour of the chromosome view of inheritance, since they strongly indicate that the factors investigated are arranged in a linear series, at least mathematically.[11]

### BRIDGES AND NON-DISJUNCTION

In 1912 Bridges found the first linkages between genes on autosomal chromosomes and the team began to build up groups of linkages so that four of them, including the sex-linked one, were discovered by 1914. There were other discoveries too: there was no crossing over in the male fly[12] and the nature of the sex-determining chromosomes was found (the male is XY as in humans). But it was two particular facts that were striking. First the number of linkage groups was precisely the number of chromosome seen in *Drosophila*. Second the explanation of the genetic consequences of non-disjunction and sex determination by C B Bridges in 1913, 1914 and 1916 in terms of the chromosome theory.

# GENES, FLIES, BOMBS...

## *THE MECHANISM OF MENDELIAN HEREDITY*

Both were seemingly inescapable indications that the chromosomes carried the genetic information and they were drawn together with all the other evidence (such as the mapping) that pointed towards this in the landmark book – *The Mechanism of Mendelian Heredity* – written by Morgan, Sturtevant, Muller and Bridges in 1915[13]. Although important sceptics remained, 1915 is often seen as the watershed after which biologists had to fight hard not to accept the chromosome theory.

Of course Morgan had embraced the chromosome theory earlier than 1915, driven by the results of his own experiments. However, William Bateson remained sceptical. His review of *Mechanism* in 1916 praised the Morgan school for its experimental work as the greatest since Mendel but was dismissive of the chromosome theory:

> The supposition that particles of chromatin, indistinguishable from each other and indeed almost homogeneous under any known test, can by their material nature confer all the properties of life surpasses the range of even the most convinced materialism.[14]

With his immediate group of British geneticists Bateson remained sceptical but his continued contact with Morgan (whom he found to have a difficult personality) led him to accept the theory in its basic form by 1921. His doubts on linkage and crossing over remained for some while and may never have been fully renounced.

A letter from his friend Clifford Dobell in 1924 told Bateson that he was planning to write on the question:

> But on general grounds, and on the cytological side, I regard Morgan and Co. with contempt. The whole thing is a colossal stunt – one of the biggest bluffs ever put up in America. I can't help thinking that if somebody, who understands their breeding experiments with Drosophila – I certainly don't – could get their real figures and analyse them properly, the whole of the evidence would evaporate away.[15]

Bateson's reply contained the following:

> We shall all look forward to your views on the chromosome question. Several times I have been on the point of doing something of the same sort, especially before my visit to New York at Xmas 1921. I am not quite easy about the wisdom of such an attack--at least if made in the spirit of your last letter. We here, who have to face the thing every day, are with you that in the crude form propounded by Morgan, it won't do. But we are equally clear that a

great part of it—probably the main part, is substantially true. We are not at all inclined to believe in the 'rosary' theory of crossing over. Nevertheless the transferable characters, almost beyond question, are somehow specially attached to the chromosomes. I don't like it, but I see no way of escape.

Bateson defended the Americans' integrity – but not much else:

They are conceited and cliquish beyond belief, but they are not out to deceive themselves or us. Their weakest point is their profound ignorance of anything but the topical and trite in genetics outside *Drosophila*, and their complete satisfaction with that ignorance.

The Danish botanist and geneticist Wilhelm Johanssen (1857-1927) was the other major critic of the chromosome theory. He had introduced the terms "genotype", "phenotype" and "gene"(in 1909). Like Bateson he resisted the theory for a decade or more after *Mechanism* but seemed eventually and rather grudgingly to become reconciled to it.

Various reasons have been suggested for Bateson's reluctance to accept the theory; a genuine belief that the evidence was not strong enough seems to have been partly responsible but the low opinion he had of Morgan (although not so much of Morgan's co-workers) certainly did not help. Cock[16] suggests that Johanssen's objections reflected a fundamentally different outlook on biology and heredity, more holistic than the particulate nature of chromosomes could accommodate.

## *MULLER'S WORK AT COLUMBIA*

Over the next three years Muller studied multiple crossing-over and the interaction between the crossings-over as it affected the chromosome mapping work. He also discovered modifier genes.

Morgan had discovered two mutants with abnormal wings. They were called truncate and beaded. The puzzles were firstly that despite intensive efforts it proved impossible to produce a homozygous truncate fly (its wings were stunted) and secondly the form the mutation took varied widely: everything from a slight notch in the wing to flies with wings shorter than their bodies. Morgan thought that this showed that mutations were unstable, the truncate gene was changing. Muller completely disagreed. Other mutations had been shown to be unchanged over many generation. Truncate and beaded could not be any different.

After much careful work – on truncate with Altenburg – in chromosome mapping of the mutations between 1912 and 1914– he found an alternative explanation in the presence of modifier genes[17]. These, while they had no apparent effect on their own, could modify the expression of the mutant gene when it was present. For truncate he and Altenburg, over the next few years, found several modifiers – indicating that things might be much more complex than Mendel had suggested.[18]

In 1914 W E Castle and J C Phillips published a paper on piebald rats which put forward Castle's long-standing conviction that genes were, in fact, unstable and fluctuating. Muller by now had enough evidence – and could assemble more from the literature – that genes were stable entities and went into print – against Morgan's advice – to join battle with the man who had inspired Morgan to use *Drosophila*. His language was blunt, suggesting that Castle had resorted to gene instability "in a spirit of mysticism". Castle responded, as was his wont, with aggressive vigour but did continue his experiments, using 50,000 rats, only to find he was wrong –and Muller was right. He retracted in print in 1919 but, as a true controversialist, did not acknowledge Muller in any way.[19] But then it was 15 years and many thousands of rats later.

In the issue of Proceedings of the National Academy of Sciences in which he recanted, Castle started a new controversy by challenging the idea – utterly central to Morgan's work – that genes were arranged linearly. Castle claimed that data obtained by C W Metz and even some of Morgan's team's own data – showed that they had to be arranged as a three-dimensional structure. He built a wire model to prove it. Sturtevant, Bridges and Morgan pointed out rather quickly that Castle had misunderstood multiple crossing over, other data came to their support and the controversy was over within a year.[20]

In his sympathetic memoir of Castle L C Dunn recalls[21] that when he told his wife that he was going to "correct" his long-held views about selection in the case of the unstable gene, she commented that he had spent a good deal of time recently in unsaying what he had said in previous years. "I agree," said Castle, "and consider that it represents progress." No such philosophical attitude is recorded when he climbed down on the 3-D gene arrangement question.

*EPISTASIS*

Muller's modifier genes were an example of what had been found by Bateson and Punnett in 1905[22], studying the genetics of the combs of chickens. They found ratios that could not be explained by simple Mendelian means. However the ratios could be accounted for if it was assumed that two separate genes were involved, one of which could mask the effects of the other on the phenotype. Bateson later[23] coined the term epistasis (meaning "standing upon") for such relationships and in the following years other examples were found. Epistasis continued to be studied into the 21ˢᵗ century as an important factor in complex diseases such as diabetes, asthma, hypertension and multiple sclerosis

## PAINTER AND GIANT CHROMOSOMES

The mapping of the *Drosophila* chromosome continued. With the stable of mutant flies providing the probes and using essentially the method invented overnight by Sturtevant, the locations of the mutated genes on the linear structure of the chromosomes was revealed. Much of the work was done by Calvin Bridges but everyone in the Fly Lab seemed to have some part. It developed an important new thread when the recently rediscovered giant polytene chromosomes in the salivary glands of the fly larva were studied by Theophilus Painter in 1933. These chromosomes are the result of the repeated replication of the DNA of the normal chromosomes without division of the results. This made them large enough to map in great detail the multitude of characteristic bands that crossed them. The work was continued by Bridges who made these banding maps and correlated them with the genetic ones derived from linkage studies. He completed it and was in the process of revision when he died in 1938. His son Phillip finished the revision which was extended over the following years.

Painter became better known outside the fly world as the subject of a legal action when he was President of the University of Texas. In 1950 in *Sweatt v. Painter* a black man, Herman Marion Sweatt, challenged in the courts the university's refusal to admit him to the Law School. The court procrastinated long enough to allow the university to set up a blacks-only law school and the matter seemed to be settled. The first appeal failed but an appeal to the Supreme Court brought a judgement that the blacks-only school had fewer facilities and, even with the laws of the time, segregation on the basis of race was illegal.

Bridges is often described as a believer in free love and a determined womaniser who abandoned his family for an exceptionally active sex life. Photographs show him as a handsome young man with a rather spectacular *bouffant* hairstyle. An orphan from an early age, he was brought up by his grandmother as a rather ragged child who did not attend school full-time until he was fourteen. He was always short of money in the early days at Columbia: he became Morgan's bottle-washer to earn some cash. He died of a heart attack at 49 but it has been suggested that this was a result of damage caused by syphilis.[24]

## LETHALS& CROSSOVER SUPPRESSORS

The other important work – perhaps mainly for the experimental opportunities it was to reveal – was on lethal genes. Lucien Cuénot had

## GENES, FLIES, BOMBS...

found unusual ratios in breeding experiments with mice and reported this in 1905. This was discussed by several authors in the following years but it was Castle and Little who showed in 1910 that the gene responsible for a yellow coat was responsible for a lethal effect: no homozygous yellow-coated mice offspring were found because they all died before birth. Similar effects were found in snapdragons by Bauer in 1908.

Morgan found the first lethal in *Drosophila* in 1912. It was sex-linked and the heterozygote showed no apparent effects. Such recessive lethals were soon found to be the largest single class of mutations occurring in the fly.[25]

A second discovery, important for later work on mutations, was the existence of crossover suppressors or modifiers. The first of these was seen by Sturtevant in 1913 (and seemed to question the model of crossing over in some degree). But, by 1919 Sturtevant had concluded that it was an intra-chromosomal effect and it was spoken of as a dominant gene. Through the 1920s there was growing evidence, from chromosome mapping, that it was an inversion[26] and, with the giant chromosomes available, this was confirmed in the 1930s.

Around 200 mutations had been classified by all *Drosophila* workers by 1919.[27]

# 4 MUTATIONS FLOURISH

## MULLER 1915

By the end of just three years full-time at Columbia Muller had managed to make important discoveries and established himself as a knowledgable interpreter of experiments and a promising theorist of genetics. He had also become embroiled in a public controversy with Castle, a senior figure in genetics – a significant rite of passage. His major achievement may well have been as a key contributor to *Mechanism*, a book generally seen as a watershed in genetics, where he had written much of the material while Morgan edited and Sturtevant was away on collecting trips. However he seems to have been less comfortable with Morgan than Sturtevant and Bridges were, finding him unadventurous scientifically. He also clearly felt that he was not getting the credit he deserved from anyone for his input to the work of the group, particularly the linkage and crossing-over work. His ideas for experiments were often taken up by the others without full acknowledgement; Sturtevant's credit to Muller in his thesis in 1915 was vague.

His ally throughout was Altenburg. Altenburg had little time for Morgan. When looking for a graduate project he had suggested that he would like to join Morgan's team and work on *Drosophila* but was mightily put out when the great man recommended the study of *Daphnia,* the water flea, instead. Altenburg promptly went to the botany department and started work on linkage in *Primula sinensis*. However, he did continue to work with Muller and the fly in his spare time. Together they planned and began a book on heredity and genetics (*Principles of Heredity*) but abandoned it when *Mechanism* came along.

## MULLER AND MUTATIONS

A chance for Muller to break away came after a visit to the Fly Lab in 1912 by the young British biologist Julian Huxley. Huxley, an Oxford don and grandson of T H Huxley, had been appointed to head the zoology department of the new Rice Institute in Houston, Texas and wanted someone from the *Drosophila* team. In 1914 he invited Muller to move to Texas and this he did in September 1915. Here, as well as teaching, Muller continued work with the flies. His main discovery while at Rice was of balanced lethals while studying the beaded wing mutation. The mutation caused the wings of the fly to shrink and develop what looked like beads around the periphery. However much it was inbred it seemed impossible to produce true-breeding homozygotes for beaded so Muller concluded that it was a dominant mutation that was a lethal recessive. This all suddenly changed when one line of beaded flies arose that did breed true giving only beaded offspring. Analysis showed that what had happened was that a quite different recessive lethal mutation had arisen on a homologous chromosome and this was so close to the position of the beaded gene that they were seldom separated by crossing over. The two recessive lethals meant that when crosses were made between flies with chromosomes of this type the only offspring that lived had the same chromosome combination. It was thus self-perpetuating.

One consequence of the existence of such chromosome configurations was that if another mutation occurred for say pink eye on the chromosome with the Bd gene then crosses between flies with this chromosome combination would not usually show up: the Bd genes would travel with the pink-eye ones with fatal results. However, if crossing over occurred then the pink-eye gene might escape from the lethal ones and sometimes make a homozygous combination that would make the fly pink-eyed. It seemed like a mutation but it was in fact a spurious one: the pink-eyed mutation may have been there for some time before it became apparent after a lucky crossing over. It seemed quite possible that a number of mutations could accumulate on a chromosome before a chance crossing over in the right place allowed them to show up in an individual. Something like this lethals could account for the dramatic sports of *Oenothera* for which the term "mutation" had first been coined by de Vries – and which had seemed to challenge Darwin's notion that evolution arose by small variations not jumps.

While Muller's achievements in Houston were by his standards not so great, the time there perhaps marks the point where he began to see mutations in a slightly different way. So far they had been tools to probe the chromosome and its mechanics; now he began to ask questions about mutations themselves. What was their nature? How did they occur? What was their significance for society, for evolution? He was not the first to ask such questions but he was the first to do so who had the expertise, knowledge and tools (the fly) to answer at least some of them.

After only a year at Houston, worried by developments in the war and perhaps pining for Oxford, Huxley returned to England leaving Muller to carry on and run the department. It meant Muller had less time than he wished for *Drosophila* work. He persuaded Altenburg to join him and undertake some of the teaching that had to be done. He had retained contact with Morgan's group, not least through time at Woods Hole, and when E B Wilson invited him to return to Columbia in 1918 he accepted. Altenburg stayed at Rice for the rest of his career but they continued to collaborate on linkage and other topics, working together at Woods Hole in the summer. For Muller the move offered the possibility of a permanent appointment[1] and would give him the time and facilities to study mutations and their generation in earnest.

## *TEMPERATURE*

Muller believed that lethal mutations were more common that any others and, with Altenburg, set out to study mutation rate on the female fly X chromosome – where a single lethal would cause death to half her male progeny. The sex ratio would be 2:1 rather than 1:1. On his departure from Texas in 1919 Muller went straight to Woods Hole and there he and Altenburg studied the mutation rate and its change with temperature using a rather refined version of the basic sex ratio technique. They found that the frequency of lethals when flies were raised at 19C (about 1%) just about doubled when the flies were raised at 27C. This was strikingly in line with van't Hoff's factor $Q_{10}$ of between 2 and 3 for the rate at which chemical reactions increase with a 10C rise in temperature and suggested that mutations were the result of chemical reactions rather than a physical event such as ionization

He and Altenburg continued to work on the problem, Muller in New York, Altenburg at Rice and together in the summer at Woods Hole. Muller fitted it in with his teaching commitments – interrupted for a day when his medical exemption from the military draft seemed likely to be withdrawn and he spent his time washing dishes only to be told that he had a heart murmur and would be spared after all. Then in December 1919 Muller learned that Morgan, Sturtevant and Bridges were to move to California for two years. Disappointed in a hope that he might be offered a permanent appointment at Columbia (he assumed it had been vetoed by Morgan), he accepted an offer from the University of Texas at Austin.

As soon as he could get away from Columbia in the summer of 1920, he and Altenburg set off for Woods Hole determined to look for other factors, besides temperature, that affected mutation rates. While doing this he developed a technique that was to allow him to measure mutation rates much more easily and it was to pave the way to his Nobel Prize.

# GENES, FLIES, BOMBS...

## ClB

The ClB method (ClB is pronounced "see-ell-bee") depends on a special structure for one of the X chromosomes of the female *Drosophila* which arose during his experiments. The C stands for a crossover inhibitor[2] (which turned out to be a large inverted section) the *l* for a recessive lethal and the B for the dominant Bar mutation of the eye. The function of the inversion is to make sure that the lethal and Bar genes stay together through meiosis. The method for detecting recessive lethals induced anywhere in the male X chromosome by some treatment or other is as follows.

A irradiated male is crossed with a ClB female and just the Bar daughters are accepted for the next stage. These will have one ClB X chromosome from the mother and one X from the father – and this may carry a lethal mutation resulting from the treatment. The Bar females are then crossed with normal (wild-type) males. All the female offspring survive. If no lethal mutation has been induced in the father's X chromosome by the treatment then half the males survive but if a lethal mutation was caused then there are no male offspring at all from the cross. Muller thus had a very clear-cut and simple way to identify induced lethals. As an additional benefit, the line of ClB females breeds true.

The head of the zoology department at Austin was John T Patterson. Patterson's speciality was armadillos; he studied the peculiar process that led one species, the nine-banded armadillo, to invariably producing identical quads. Patterson was also a good administrator of research and he made sure that Muller had enough resources (even an assistant) to carry on with the fly work. The two other members of the department were Carl Hartman, a physiologist, and Theophilus Painter, an adjunct professor with an interest in lizard cytology who, as we have seen, worked with the giant polytene chromosomes.

The first year at Austin was disappointing. Although he now had the time and resources to do his fly research, the ClB technique threw up a serious problem. Whereas with the sex ratio method he had found an extraordinary 1% mutation rate for lethals on the X chromosome , with the ClB technique he found just a single mutation in nearly 3000 gametes examined. Even worse, the spontaneous mutation rate varied wildly between fly stocks and he could see no appreciable effects from ageing or application of ether. The chance of seeing again with ClB the temperature effects he and Altenburg had found using the sex ratio method seemed remote.

## MUTATIONS FLOURISH

If the particular line of research was not going well Muller kept busy. He developed his general ideas about the gene, saw Bateson's public acceptance of the chromosome theory, and travelled to Europe and Russia.

### MULLER IN EUROPE 1922

Illustration 5: Muller and Altenburg 1922

He crossed to England in June 1922 on the *Aquitania* with Altenburg, arriving 19 June, where they met R C Punnett. They then flew to Paris (where Muller had his photograph taken at the top of the Eiffel Tower) and Muller went on to Germany and then flew to Moscow (a nine hour flight in those days) where he spent three weeks. He visited the Institute of Experimental Biology in Moscow, where he left cultures of *Drosophila* with A S Serebrovsky. He then went to Petrograd[3] to meet N I Vavilov at the Institute of Applied Biology.[4]

Morgan and Sturtevant were also in Europe that year. Sturtevant and his new wife Phoebe (one of Morgan's assistants) arrived on their honeymoon (which was to be taken in England, Norway, Sweden and Holland) on the 1st May to be followed on the 26th May by Morgan (whose initial address in the UK was given as the John Innes Institute). Morgan delivered the Croonian Lecture – *On the Mechanism of Heredity* – at the Royal Society in London on 1 June.

### MULLER MARRIES

On 11 June 1923 Muller married Jessie Marie Jacobs. A graduate of MacPherson College and an early female PhD (thesis title *The Trilinear Binary Form as a Cubic Surface*) of the University of Illinois, Jessie started lecturing in pure mathematics (she is described as an "algebrist") at Austin in 1920. An active professional, she was editor of the Texas Mathematics Teachers Bulletin. She and Muller seem to have met when he asked her for help with some statistics. They honeymooned hiking and camping in the National Parks.

In early 1924 Jessie became pregnant and David Eugene Muller was born on 2 November. On the grounds that an academic career and a baby didn't mix, the mathematics department terminated her appointment. However, she did carry on working with her husband, collaborating on a paper examining and tidying up the statistics associated with making chromosome maps from crossing-over experiments[5] Jessie is honoured at the University of Illinois through their lecture series *Distinguished Women in Mathematics*.

## TWINS

About this time Muller undertook IQ studies on a pair of twin girls. These may well have been the first such studies to be conducted on a sound scientific basis.

The two girls had been separated shortly after birth and raised separately until they were 18 years old in somewhat different backgrounds. One was brought up in mining camps; the other on a dairy farm. One had little schooling after the age of 12, the other went through high school. One became a secretary, the other a teacher.[6]

Muller first had to establish that they were, in fact, identical twins and compared almost a dozen physical attributes to be confident that they were. He then went on to test their IQs and found them to be very similarly high, putting both of them in the top 2% of the population. This was unlikely to be a coincidence. It possibly implied a large genetic influence but the backgrounds they were brought up in were similar so it was impossible to draw such a sweeping conclusion. If the twins had had very different upbringings then the result would have been more convincing  Muller thought more research was needed and he had certainly established some procedural principles that had to be followed if such work was to stand up to scrutiny. He wrote two papers about the twins[7] but didn't follow up the study himself. His acknowledgements in the first paper included Jessie:

> The writer wishes here to acknowledge ... the efficient and invaluable cooperation rendered by both of the twins, and by the writer's wife, Dr. Jessie Jacobs-Muller.

Jessie's involvement may well be reflected in the well-developed methodology – which was, of course, the enduring value of the paper.

## MULLER AND RADIATION

Muller started to use radiation as a tool for genetic work in 1923. He used it to "weaken" cells and so to expose genetic deficiencies and studied its effect on crossing over frequency. In all these studies, he did not test irradiated flies for mutations even though he had the ClB technique ready at hand. Carlson has suggested that this was because his mind was set against radiation as an effective mutagen as result of his earlier experience with the $Q_{10}$ work which led him to believe that spontaneous mutation was a chemical process. It was not until late 1924 that he thought of looking at radiation effects and mutations but even then the planned project with Altenburg did not go ahead and nothing more happened until early 1926. This was when Muller reviewed the literature on radiation and mutations. He saw that Morgan, Payne, Harold Plough and others had used radium and x-rays without any conclusive result[8]. But other experiments showed that radiation did have dramatic effects on cells. Bardeen in 1906 had exposed the sperm in toads to x-rays and found that the embryos they fathered with unirradiated mothers developed abnormalities, showing that the radiation damaged the nuclear material. The results of Paula and Oscar Hertwig were interesting. They had found that sperm exposed to moderate doses of radiation gave rise to abnormal embryos but if the dose was increased some embryos survived longer. At the higher dose the embryos were characteristic of those produced parthenogenetically. It became called the Hertwig effect and was yet another indication that radiation had some profound effect upon the genes. Altogether the laboratory results, combined perhaps with the growing evidence that radiation had had disastrous health consequences for some of the physicians who had used it, appear to have convinced Muller that the experiments he had avoided for so long[9] might be worthwhile.

Muller initially conducted three experiments in November 1926 in which he irradiated batches of male and female flies with different doses of x-rays and then searched for mutations. In the first he used a variation of the sex-ratio technique he had developed in 1919 with four different irradiation times and immediately found striking results. The mutations increased regularly in number depending upon how long the flies had been irradiated and they could be traced back to females where just females had been irradiated and to males where they had been the sex exposed. He found a similar result using his ClB method. Both experiments showed an increase in visible mutations, as well as the lethal ones, with dose and these seemed no different from the

# GENES, FLIES, BOMBS...

spontaneous mutations that had been studied for so long. A third experiment producing just visible mutants was consistent with the other two and even suggested[10] a doublet nature for the gene.

Muller wrote a rather coy paper for *Science*[11] in which he outlined his results in just four pages without any details of the experimental methods used. The language was not quite the dispassionate tone of scientific publications: "a rise of about fifteen thousand per cent. in the mutation rate over that in the untreated germ cells." But in the short publication he did point out that he had produced a large number of mutations, some quite new and some that seemed very like familiar ones that had arisen spontaneously, and that most of them were stable through, at least, several generations. He noted, characteristically, a few practical factors relevant to humans. Doses used to sterilise people could have long term effects after the irradiation had stopped and when fertility returned. He also saw links with both the known greater radiation sensitivity of dividing cells and the fact that radiation occasionally caused cancer. He closed the paper with remarks on how valuable it would now be to be able to produce mutations to order; it would permit "respectable" genetic mapping in other species and might be of use to "practical breeders". However, while the paper was quite wide-ranging, the failure to give any details of methods led to it being treated with suspicion. Morgan was shocked and deeply sceptical about the 15,000 percent increase. He said: "Now he's done it. He's hung himself."[12]

But of course Muller had not. He had been invited to give an address at the Fifth International Congress of Genetics held in Berlin in September 1927 and he used the opportunity to produce a paper entitled The Problem of Genic Modification.[13] In this twenty six page paper, prepared on the voyage to Europe and revised right up to the moment of presentation, Muller set out in detail the carefully-planned experimental strategy and procedures and the results he had obtained. The paper caused a sensation: man could modify his precious genetic material at will and even had the power to "speed up evolution". Muller went to Berlin a fairly obscure scientist but returned to the USA a celebrity.

## *STADLER*

Muller presented his results at a meeting of the American Association for the Advancement of Science in Nashville, Tennessee that December. Another presentation at the meeting was by Lewis J Stadler from the University of Missouri. Stadler had also found the mutagenic effects of radiation, in his case, in barley. He had irradiated barley and maize seed and planted it in 1926 and planted the offspring in the summer of the following year. From the 2200 or so barley seedlings derived from irradiated parents he found seventeen mutations, from 1300 controls none. So, it was a quite convincing result and if Stadler had chosen another experimental subject – one with a cycle time of less than a year – he might have published before Muller or at least had more of the public credit for his independent work. In fact Stadler's work was not as complete as Muller's – not least because Muller had the benefit of a subject which had been intensively studied and characterised. Stadler accepted his misfortune with good grace and even friendliness and Muller was generous in his acknowledgement of Stadler's work.

Within just a year or two of Muller's Berlin address the mutagenic effect of radiation was confirmed in *Drosophila* by Weinstein, in *Datura* by Blakeslee, in tobacco by Goodspeed and in the parasitic wasp *Habrobracon juglandis* by Whiting.[14]

Muller himself took on new students and his work spread in many directions. They investigated the possible mutagenic effects of chemicals with negative results, extended the radiation work to include radium and devised a way to test if ultraviolet radiation caused mutations (it did). With Painter, Muller used radiation-induced translocations to construct a cytological map of the chromosome. It confirmed the order of mutations along the chromosome as in the genetic map but showed that they were not in the physical locations previously thought. Muller had thought from the start that the radiation-induced mutations were of the same nature as spontaneous ones. He had some success in showing that he could induce reverse mutations with radiation, returning the gene to its state before the original mutation. It became an important issue because Stadler had become convinced from his further experiments that what he and Muller were dealing with were not mutations of individual genes but more dramatic disruptions (deletions) of chromosome structure. A demonstration of reversibility made this much less likely. In 1929 Muller considered the possibility that spontaneous mutations were in fact caused by background radiation. With Mott Smith he showed that this was very unlikely.

## *DOSAGE COMPENSATION*

In 1931 he became interested in dosage compensation. With two X chromosomes, female flies have two copies of the genes located on it while males have only one. For some characteristics determined by the X chromosome there seemed to be compensation mechanisms for this but for others not. A clever series of experiments showed that the matter was complex and other matters pressed in preventing him investigating further.

# GENES, FLIES, BOMBS...

While he continued with his work, Muller's world began to fragment. His relationships with Patterson and Painter deteriorated and there were persistent criticisms of the very basis of his radiation work from Stadler and the Swede Heribert-Nilssen. Nilssen claimed that Muller had simply destroyed normal cells with radiation, allowing existing mutant ones to flourish. Personally too things had gone wrong. His marriage to Jessie had broken down under the strain of his workload and the loss of her job as a mathematics lecturer when she became pregnant.

## *AGOL, LEVIT, RAFFEL, OFFERMANN*

In 1931 Muller was joined by two Russians: Israel Agol and Solomon Levit. Both were on Rockefeller Fellowships. Agol, initially, as many other Russians was a Lamarckist who had switched from philosophy to biology ad worked in Serebrovky's laboratory. He had helped in Serebrovsky's work confirming Muller's discovery of the mutagenic effect of radiation which removed the objection to gene theory, for a Marxist, that it was an immutable, abstract and immortal concept. He had some success in novel gene mapping work in *Drosophila* and the Fellowship seemed a wonderful opportunity to study with Muller in Texas. He wrote to Muller in 1930 and in 1931 arrived in Austin.

A Party activist since the Revolution and a fighter for the Red Army during the civil war he was well-versed in the Marxist philosophy which "permeated his organisation, scholarly and scientific work." In the early 1930 there was a dispute between two wings of Party: the "dialecticians" against the "mechanists". The dialecticians (led by the philosopher Deborin) believed that progress was achieved in all fields through the process elaborated by the German philosopher Hegel in which an idea and its opposite clashed to produce a synthesis different from either. The mechanists believed that while this might be true for social ideas, the progress of science was better described by different dynamics. Agol was a disciple of Deborin[15]. This seemingly academic affiliation, in a dispute between (it might be thought) incomprehensible and possibly meaningless propositions was to prove fatal for him.

For the time being he worked with Muller on *Drosophila* with enough success to encourage Serebrovsky to encourage one of his former students, Solomon Levit, to join them. Levit had followed a similar path to Agol and was a convert from Lamarckism after Muller's discovery. A committed communist he seems not to have quite had Agol's philosophical background. He too settled to a period with the master during 1931 with a particular interest in Muller's twin studies.

## MUTATIONS FLOURISH

The team was made up with four others, two of them were to play a longer-term part in Muller's life. Carlos Offermann was an Argentinian with a German background from the University of Buenos Aries, a slight man and according to Muller, a "hunchback, but a fine, intelligent chap". He was to play an unexpected role in Muller's private life. He arrived in Austin in late 1930. Daniel Raffel with a PhD from Johns Hopkins, and a background in the genetics of *Paramecium* came soon after.

### *SUICIDE ATTEMPT*

On 10 January 1932 he wrote a letter to Altenburg putting his academic affairs in order (but did not send it) and set off into the hills around Austin, where he swallowed a handful of sleeping pills. He was found the following day by a search-party of his students, dazed and muddied but alive.[16]

He returned to work apparently as enthusiastic as ever but clearly welcomed the award of a Guggenheim fellowship. This would take him to Berlin to work with Timoféeff-Ressovsky and he had already decided to leave Austin, his working relationship with Patterson and Painter at an end. He and Jessie decided that she would not come to Berlin, a trial separation might help them solve their problems. After Berlin there was the prospect of something quite new: a long-term appointment in the USSR.

### *THE SPARK*

Before he left there were to be two dramatic acts that revealed just how profoundly disaffected he had become. He had helped form the University of Texas branch of the left-wing National Student League[17] at Austin and the first act was to sponsor the newspaper, *The Spark,* published by them. He wrote some of the articles and persuaded Altenburg to distribute the paper at Rice.[18] The headlines of the first (and only) edition distributed in June 1932 included:

UNIVERSITY WORKERS DEFRAUDED

Students and Workers! Form a United Front!

WAR!

College Students and Workers Will be Shot Down to Protect Capitalist Investments!

STARVATION IN AUSTIN

Results of Student Survey Reveal Appalling Conditions

# GENES, FLIES, BOMBS...

There were also articles about the Scottsboro Boys and the struggle of the miners of Harlan and Bell Counties in Kentucky. The Scottsboro Boys were nine African-Americans accused of raping two white girls on a train in Alabama in 1931. They were initially found guilty and facing death sentences but over several trials and appeals (where the defendants were supported by the American Communist Party) the sentences were commuted to long imprisonments. Four of the Boys were eventually acquitted but most of them spent long years in US jails. The Harlan County War as it is sometimes called was a series of strikes in Kentucky organised by unionised miners protesting first about wage cuts and later against sackings of protestors. The War, which lasted from 1931 to 1939, was punctuated by episodes of extreme violence on both sides including shooting and bombings. The violence spread to neighbouring Bell County.

The National Student League had been penetrated by the FBI and they had Muller under surveillance. It may have been the beginning of a file they kept on him that eventually reached, according to Carlson, some 500 pages[19]. The FBI reported on Muller's activity to the President of the University. Given that the newspaper was strictly in violation of university rules and an extraordinarily inflammatory publication for a senior member of staff to support, the University must have looked forward to the departure of their celebrity faculty member.

Illustration 6: *The Spark* cartoon

Muller's second act was hardly less dramatic, dealing a damaging blow to a movement he had supported since the beginning: he spoke at the 3rd International Eugenics Congress at the American Museum of Natural History in New York in August. This aspect of his life is dealt with in a later chapter.

## MORPHS ETC

Muller went straight from the Eugenics Congress to present a paper to the Sixth International Congress of Genetics at Cornell University, Ithaca. The paper was titled *Further Studies on the Nature and Causes of Gene Mutations* and suggested indirect action as the principal cause of radiation-induced mutations and presented a detailed and novel mutation classification. This was made on the basis of the effects on function. Some mutations reduced gene function and they were either a*morphs* – when the gene function was completely lost as a result of the mutation – or *hypomorphs* – where there was a partial loss. These were generally recessive (although a hypomorphic mutation could be dominant if there was haploinsufficiency).[20] Dominant alleles were classified as *hypermorphs* if they increased gene function, *antimorphs* if they acted against normal function and *neomorphs* if they resulted in a new function, different from the normal one. The classification clarified possible gene actions considerably and is still in use today among *Drosophila* geneticists.

## MULLER LEAVES USA

Interesting and important as this was, it was Muller's behaviour that attracted much attention: clearly nervous, he struggled with his typically scrappy notes to give an distracted and disjointed presentation. T H Morgan observed to a colleague after the address that "Something is wrong with Muller."[21]. Muller left New York for Europe just a few days later on 5 September 1932 on the German luxury liner *General von Steuben*[22] with a Guggenheim fellowship and a plan for a sabbatical.

## POSITION EFFECT

Morgan, Sturtevant and Bridges had been very active since the departure of Muller. Sturtevant made a major discovery – the position effect – and Bridges' cytological skills helped explain it. It came from a study of the Bar eye mutation, the one Muller had used in his C*l*B technique.

The Bar mutation is a sex-linked dominant one that reduces the size of the eye, making it narrower and reducing the number of facets (ommatidia) that make up the compound eye. There had been some puzzles about the pattern of its inheritance, notably the reversion of very occasional offspring to normal eyes. This was studied by Zeleny around 1920 and he concluded that the reversion occurred in females late in the development of the eggs. He also found a more severe form of the mutation which he called ultra-Bar, renamed double-Bar by Sturtevant in what turned out an ill-advised move.

Sturtevant created females with a Bar mutation on each homologous (X) chromosome[23] but one of these had marker mutations on either side of the Bar. He used forked (f) and fused (fu) as the markers. He found that the reversions were all the result of cross overs occurring between the marker mutations. The results he saw led him to believe that unequal crossover was occurring leading to two Bars on one chromosome and none on the other. Hence ultra-Bar and wild type. But this was not all. It might have been expected that the eye of a homozygous Bar would have the same number of facets as one with the double-Bar and wild mutations. After all, each had a dose of two Bars. This was not so; there were marked differences for this combination and others where the gene numbers were the same but their locations were different. Sturtevant found a similar result for a mutation infra-Bar, a partial reversion towards normal eye. Until then it had been thought that genes acted independently of their position. Now it was clearly not so. This was a wonderful demonstration of the power of the meticulous methods developed by the group, showing just how far the nature of the chromosome could be penetrated by carefully prepared crosses and fly counting.

Direct support would came from Bridges and, quite independently, Muller[24] in the next decade. (Muller lays out quite clearly his claim to at least joint recognition and possibly priority). They showed that, in the giant polytene chromosomes of the salivary glands, Bar itself was a duplication of a small section of the X chromosome and ultra-Bar – or double-Bar as Sturtevant called it – was in fact a triplication.

In 1928 Morgan, anticipating retirement at Columbia, moved, with Sturtevant and Bridges, to the California Institute of Technology as Head of the Division of Biology.

### *HETEROSIS*

The progress made in understanding fundamental questions with the fly and other systems was extraordinary in these early decades. However, it was academic and had little direct practical application to stock breeding or crop improvement. Heterosis was one phenomenon that it threw just a dim light on.

## MUTATIONS FLOURISH

Until the early years of the 20th century the main way in which breeders developed stocks of plants and animals with the particular characteristics they wanted was mass selection. The method was an obvious and ancient one: collect the seeds from the plants that tended towards the ideal sought, plant them and repeat the process over many generations. Mate the best animals. Eventually (and it could be a slow process) you would have plants that gave a larger yield, pigeons that flew faster.

Hybridisation, in the sense of crosses between two strains of plant in an attempt to produce plants with more desirable characteristics, had been practised since the seventeenth century[25] and it had been recognised that there were two results: inbreeding usually produced weaker stocks with more defects (in-breeding decline) while hybridisation usually produced more vigorous stock. The latter effect became called hybrid vigour or heterosis.

Explanations for this effect could hardly be convincing before Mendel's genetic ideas became accessible after their rediscovery and it was not until 1908 that George Shull, at Cold Harbor Spring, suggested a plausible mechanism for hybrid vigour in maize. He had taken inbred lines – which showed deterioration in yield and vigour – and crossed them. There was an immediate recovery with some of the hybrids having a higher yield than the original stock used in the inbreeding programme. His explanation was based on the genome of a hybrid being likely to contain more heterozygotes (more loci with dissimilar alleles) than the original stocks. If the original stocks were in-bred then they would contain a high proportion of homozygotes and the results of crossing them would produce a dramatic increase in vigour. This was what was seen. As Shull stated his hypothesis in 1914 (when he gave the effect the name "heterosis"):

> The physiological vigor of an organism as manifested in its rapidity of growth, its height and general robustness, is positively correlated with the degree of dissimilarity in the gametes by whose union the organism was formed. The more numerous the differences between the uniting gametes - at least within certain limits - the greater on the whole is the amount of stimulation.[26]

Edward East, at the Connecticut Agricultural Experiment Station, had found the inbreeding effect at the same time but had not tried the hybridisation so had not found heterosis. His work and he himself supported Shull's conclusions.

Hybrids came to dominate the maize planted in the USA. Initially there were difficulties because the in-bred strains that produced the seeds had a very low yield but this was much improved after a suggestion by Donald F Jones, one of East's students, that four-way crosses should be undertaken. These involved crossing two pure lines and then crossing the hybrid resulting with another hybrid from two other pure lines. It was a welcome development. Mass selection was not significantly improving corn yield and the hybrids came to dominate the corn planting. In Iowa while just 10% of corn was hybrid in 1935 by 1939 it was 90%. [27]

While heterosis was an established fact, the explanation of it was, at various times, controversial. The alternative explanation to that of Shull and East was the so-called "dominance" hypothesis suggested by Davenport in 1908 and developed in 1910 by Bruce and independently by Keeble and Pellew.

It explained in-breeding decline by suggesting that it led to normally-recessive slightly detrimental alleles being present at multiple loci in both parents and and thus being expressed. Hybridisation introduced normal dominant alleles back into the genome and these then masked the recessive alleles. The dominance hypothesis quickly became the favoured one and persisted until the 1940s when Fred Hull and others pointed out that it seemed unable to account for several observations, notably the failure to get significant improvement by mass selection and by inbreeding. This caused a return to favour of the earlier hypothesis – now generally known as "overdominance" – for about a decade. It was now couched much more firmly in Mendelian terms and had the advantage of several decades of refinement in genetics. However, by the end of the 1950s opinion swayed back towards dominance as an explanation as better experimental management did show improvements from mass selection. Crow, who had for a while been an enthusiastic advocate of overdominance could say in 1998 that:

> ...the dominance hypothesis is the major explanation of inbreeding decline and the high yield of hybrids. There is little statistical evidence for contributions from overdominance and epistasis.[28]

Two paragraphs from a recent review[29] suggest that we have not heard the last words in this long debate:

> Although the current view favors dominant allelic action as the major contributor to heterosis (Coors and Pandey 1999; Crow 1999), there are several lines of evidence that suggest that mechanisms beyond simple complementation may be important

## MUTATIONS FLOURISH

in heterosis.

The dominance/overdominance debate becomes even more nuanced when contributions of epistasis are considered. Epistasis is classically defined as interactions between genes at two (or more) loci affecting the phenotypic expression of a trait. With regard to heterosis, a dominant-acting locus may interact with an overdominant-acting locus to further heterotic gain. Therefore,the genetic background and allelic interactions therein can have an effect on the heterotic contributions of individual loci

# 5 EVOLUTION AND EXILE

## *POPULATION GENETICS AND EVOLUTION*

The first person to apply mathematical techniques to Mendelian inheritance was Mendel himself with his calculation of the consequences of repeated self-fertilisation of his peas. The more important case of Mendelian inheritance in a randomly mating population where there were just two possible alleles of a gene was looked at by Yule, Castle and Pearson from 1902 onwards but no general solution was found until 1908 when the pure mathematician G H Hardy and Wilhelm Weinberg, a German physician, arrived at it independently. Hardy quickly solved the problem and published his solution in June 1908.[1] It became known as Hardy's Law which was rather unfair to Weinberg since he had published the solution six months earlier as part of a study of identical twins. Weinberg also anticipated some of Fisher's conclusions described below but was ignored because his approach was too mathematical for many geneticists and, perhaps, because he published in German[2].

From around 1912 there were mathematical studies in the USA of inbreeding which seemed important to fixing good characteristics in livestock breeding and yet clearly had degrading effects. An influential publication in the UK was R C Punnett's *Mimicry in Butterflies* published in 1915. Doubtful that mimicry[3] could evolve by small steps, he commissioned a Cambridge mathematician, H T J Norton, to calculate the effects of selection on gene frequencies. Norton's results showed that natural selection could happen much more quickly than expected. (Provine again)

In 1918 the British geneticist Ronald Aylmer Fisher produced a paper that resolved one of the issues between the Mendelians and the biometricians. The biometricians were convinced that evolution could proceed only through small changes – as Darwin had proposed – and had thus studied continuously varying quantities like height. Such studies led fairly readily to a blending theory of inheritance. Mendelism, with its discrete and distinct characters, seemed at odds with this. It was Fisher who showed that the inheritance of seemingly continuous characteristics could be explained as a result of a number of genes acting together, each inherited in a Mendelian fashion. It was not a new idea (Ulny had offered it in 1902 as was noted earlier) but Fisher's brilliance in mathematical analysis resulted in a paper that has been seen as the beginning of modern theoretical population genetics. It was also not published without acrimony. Originally submitted to the Royal Society of London it was refereed by Pearson and Punnett who expressed reservations about it; after enough disagreement to produce lasting bad feeling between Fisher and Pearson, Fisher sent it to the Royal Society of Edinburgh who published it in their Transactions.

This, of course, was very relevant to ideas of evolution and over the next decade Fisher developed the analysis to show that natural selection could cause changes in gene frequency in a way that made Darwin's original idea of evolution through selection acting on small changes inherited in a Mendelian way at least plausible. Mutations, provided they were not eliminated by chance early on, could spread through a population and become the new and fitter norm. Saltations were unnecessary as was the inheritance of acquired characteristics and orthogenesis. Indeed Fisher put forward a simple but persuasive geometrical argument that small mutations were more likely to increase fitness than large ones were; a phenotype would be near to a fitness optimum and a small mutation would have a 50/50 chance of pushing it even closer. A large mutation would be much more likely to drive it far away, perhaps leaping right over the optimum.[4]

He drew his ideas together in *The Genetical Theory of Natural Selection* published in 1930. Marred slightly to modern eyes by the large fraction devoted to eugenic theories, it has been called the greatest work on evolution since Darwin.

Fisher was also the outstanding statistician of his age and in this, as in the genetical theory, he was celebrated for his obscurity in mathematical analysis. He had poor eyesight all his life and learned his mathematics without writing much down. It led him to leave out explanatory steps essential to lesser mortals.

J B S Haldane developed population genetics in a series of publications over 10 years but his major contribution may have been his book *The Causes of Evolution* where he put forward the key ideas in non-mathematical terms in the body of the book and summed up the theory in a mathematical appendix. Haldane was a skilled populariser of science and the book made the new ideas more accessible to many readers. One of Haldane's technical achievements was to show through examining natural selection in action, that it could work more quickly than Fisher envisaged[5].

One of the examples in his first paper[6] was the Peppered Moth (*Biston betularia*). This moth exists in three forms: a pale one, a much darker, melanic one and a rare intermediate. In Manchester in 1848 the commoner form (99%) was the pale one but by 1901 over 99% of the moths were melanic. Haldane calculated in 1924 that the dominant melanic had been selected over the recessive pale in the ratio of 3:2. Such intense selection seemed improbable at the time but in the 1950s Kettlewell showed that it could be explained by the moth's ability to hide from predatory birds by pressing itself against tree trunks: light moths were more likely to survive in unpolluted areas and dark ones when the trunks were blackened by pollution[7].

Sewall Wright also saw Mendelian inheritance and natural selection as the bases for evolution but added two additional factor: interacting genes and genetic drift. Wright had an early interest in the inheritance of colour combination in guinea pigs and encountered the interacting systems of genes that determined this. When he joined the US Department of Agriculture in 1915 he not only extended the colour inheritance studies to other mammals but also became involved in inbreeding studies on guinea pigs that had been running for almost ten years[8]. Applying an analysis method (path coefficients) he had developed earlier, he devised breeding programmes based on inbreeding families and taking advantage of the interacting genes. When he moved to academia he looked at how what he had found might relate to evolution. So he focussed on small, isolated populations and the effects on them of the inevitable inbreeding that occurred. This would result, because of the small numbers involved, in the drift of the gene structure of the group away from its base. Then, changes in combinations of interacting genes, could take the whole population away from the adaptive peak it had been at. Natural selection could then take it to another one.

Although there were differences in emphasis and disagreements (the relevant population size, the role of genetic drift and the evolution of dominance) the three men had in common a belief that Mendelian genetics had been reconciled with evolution by natural selection. The question of the origin of species was left largely unanswered (although Wright had shown a way this might be resolved) but it was a remarkable synthesis in the early 1930s of two ideas that had been widely thought to be quite incompatible. It became known as the Modern Synthesis after the title of Julian Huxley's 1942 book *Evolution: the Modern Synthesis*.

Fisher and Wright famously disagreed over the evolution of dominance although this was a reflection of differences in the general view of evolution[9]. It was recognised that mutations were overwhelmingly recessive to the wild type and Fisher thought that his had to be the result of natural selection acting on (unidentified) genes that somehow modified the effect of the mutations so that while initially co-dominant they became recessive over time. Sewall Wright had a quite different explanation. Each wild-type gene was responsible for producing more of the enzymes (seen as the principal role of genes) than were needed by the body so if a mutation occurred that resulted in it not being fully functional, it would have little or no effect. The copy on the homologous chromosome would be adequate. Wright's explanation proved to be nearer to the right one. The debate between them over this and other issues of evolutionary theory was conducted in the public literature between 1930 and 1962 and became increasingly sharp, with accusations of misrepresentation from both sides. The evidence that dominance was simply a result of the nature of biochemical processes increased and Fisher's ideas were overtaken: natural selection was not needed, dominance was intrinsic to the system. It was rather ironic that it later became clear that natural selection may well have had a part in producing genes with overcapacity since it quite probably would give their host some advantage through robustness to mutation. There was still a debate going on into the 21st century[10].

Sergei Chetverikov headed a research group at Nicolai Kol'tsov's Institute of Experimental Biology. His 1926 paper drew together naturalists, geneticists and biometricians in a synthesis[11]. Available only in Russian the paper was hardly known outside the USSR and Chetverikov's contribution was hardly known outside the USSR until much later.[12]

> His work with *Drosophila* was one of the first demonstrations of the widespread genetic variation in natural populations, and his influence on N. Timofeev-Ressovsky, B. L. Astaurov, N. P. Dubinin, and, to a lesser extent, Theodosious Dobzhansky established the foundation for a strong program in evolutionary genetics.[13]

A major milestone in understanding speciation was Theodosius Dobzhansky's *Genetics and Origin of Species* published in 1937[14]. His essential thesis was that, as had already been suggested, new species originated from populations that had become geographically isolated. The species then became better adapted to the local conditions through natural selection working on the genetic variation already existing in the population. It made intra-population genetic variation and polymorphism central to the whole issue of speciation. This challenged the classical genetic thinking of Morgan and Muller because their view was that there was in fact rather little variation: most individuals were homozygous for what they termed the wild type gene. Variations for them were those gross variations that showed up as mutations in their experiments.

In 1937 Dobzhansky could offer little evidence for variation: human blood groups, polymorphism in *Drosophila* and a few other studies. By the third edition of the book in 1951 there was much more, if not yet at the individual locus level needed to underpin his theory.

Starting with Fisher, Haldane and Wright the simpler assumptions of the Hardy-Weinberg relation were steadily removed so that different mating patterns were studied, mutations and multiple alleles were included and migration was accounted for. Evolution continued to be a major focus of population genetics for the remainder of the century but, at least as important for the understanding of the consequences of mutations, it provided a mathematical framework to model their transmission and some measures of their likely impact on populations.

One consequence was that the role of mutations in evolution was changed. The view of, for example, Morgan that mutations (and of course he had rather drastic ones in mind) somehow were the trigger for evolutionary change faded to be replaced by one where mutations were the ultimate source of the variations within populations. The new view was that variation – which was after all stored mutations that had not been selected out and shuffled by recombination – provided enough material for natural selection to work on as soon as the challenge arose.

The persistent criticism of population genetics was that it failed to represent the full complexity of the way inheritance worked with most notably the comments of Mayr that it dealt only with individual genes – he called it beanbag genetics because it was as if the genes were plucked individually from a bag. It was an essentially reductionist approach that took little account of the complex interactions between genes that required thinking at the level of the whole genome.

Without mutation the variations in genome would be shuffled by sexual reproduction but no new cards would be added. Artificial selection shows that considerable changes can be achieved just through exploiting these variations. Mutations increase the variation present in the population provided the mutations survive. This can happen even if they are slightly deleterious through protection by dominant alleles. Fisher thought it needed large populations to sustain mutational variation. Wright thought variations could be sustained (and amplified) in small populations by genetic drift.

*MULLER IN GERMANY*

Muller used the crossing to Europe to study communist theory. He read Lenin with a plan in mind to acquaint himself with all the key Marxist writings. After visiting his sister in Munich he arrived at the the Kaiser Wilhelm Institute for Brain Research in Berlin-Buch on 9 November 1932. Here he expected to work with N V Timofèeff-Ressovsky, the Russian who had been working on spontaneous mutations in *Drosophila* and who had quickly confirmed Muller's demonstration in 1927 that radiation caused mutations. They planned to make more detailed experiments on the effectiveness of x-rays of different wavelength and both had an interest in the physical processes that caused chromosome damage.

Muller was present at one of the most significant events in 20th century science during his stay. He went to Copenhagen and met Niels Bohr in the spring of 1933. Bohr had given his lecture "Light and Life" in August of the previous year[15] and in it had suggested that the laws of life were quite different from the laws of physics and chemistry and certainly not reducible to them. It was like the complementary wave and particle explanations of quantum phenomena. Muller found Bohr's views on biology "hopelessly vitalistic"[16]. However the lecture had been attended by the theoretical physicist Max Delbruck and he became fascinated by the prospect of finding these other laws[17].

When he moved to Berlin to work with Lise Meitner later in 1932, Delbruck joined Timofèeff-Ressovsky and K G Zimmer in regular discussions on possible physical and chemical models of genes. It had become clear from Muller and Timofèeff-Ressovsky's work that mutations were probably caused by a single ionisation event or a small cluster of events – pointing towards genes behaving like molecules. The discussions deepened into collaboration after Muller left Berlin and the result was a paper[18] (the *Three Man Paper*) which is often seen as the first stirrings of molecular biology (and thus the beginnings of a science which was to revolutionise genetics). They summarised their conclusions as:

> Consequently, we view the gene as an assemblage of atoms within which a mutation can proceed as a rearrangement of atoms or a dissociation of bonds (triggered by thermal fluctuations or external infusion of energy), and which is largely autonomous in its operations and in relation to other genes."[19]

As Delbruck said in his Nobel Lecture address in 1969:

> Genes at that time were algebraic units of the combinatorial science of genetics and it was anything but clear that these units were molecules analyzable in terms of structural chemistry. They could have turned out to be submicroscopic steady state systems, or they could have turned out to be something unanalyzable in terms of chemistry, as first suggested by Bohr – and discussed by me in a lecture twenty years ago. It is true that our hope at that time to get at the chemical nature of the gene by means of radiation genetics never materialized.

# GENES, FLIES, BOMBS...

Muller had not arrived in Berlin at a settled time. Hitler had become Chancellor at the end of January 1933. In April the official persecution of the Jews began, trades unions were soon outlawed and in July political parties, except the Nazis, were outlawed. Universities were always suspected of harbouring communists and other anti-nazis. Oskar Vogt, the head of the Institute, refused to denounce Jews and anti-nazis and, as a result, the Institute was invaded by stormtroopers in March, Vogt's house was broken into and several staff were assaulted and arrested. Muller managed to escape. Soon many academics and others were leaving Germany and it became clear to Muller that he had no future as a known leftist. Given the brutal eugenics programme Hitler had in mind there was little scope for a principled geneticist either. It was time to leave. Return to the USA was not an attractive prospect: relations with Patterson and Painter in Texas were poor so he was disinclined to go there but employment for academics was hard to find elsewhere.

Fortunately, he had already had an offer from N I Vavilov who had visited Muller in Berlin (and had sponsored him as a corresponding member of the Soviet Academy of Sciences) : Muller was invited to work in the genetics laboratory in Vavilov's All-Union Institute of Plant Industry[20] (called the Institute of Applied Botany until 1930) in Leningrad (now St Petersburg). In September 1933 he was joined in Berlin by Jessie and David and on 16 September they set off for Leningrad. Muller took a suitcase of flies, food for them, 10,000 glass vials and 1,000 bottles. Other belongings, including a 1932 Ford car, were to follow.[21]

## BRIDGES IN RUSSIA

They must have met Bridges in Leningrad. Having spent the period October 1931 to May 1932 lecturing in the USSR, Bridges returned for the autumn of 1933, largely to work with Alexandra Prokovfaya-Bergovskaya.

## TIMOFÈEFF-RESSOVSKY

Nikolai Vladimirovich Timofèeff-Ressovsky was born in 1900 and had his education as a biologist at Moscow University interrupted by the Bolshevik Revolution of 1917. He volunteered to fight first the Germans in the infantry and cavalry and then, after the Bosheviks signed the peace treaty of Brest-Litovsk with Germany in 1918, he joined the Red Army (in some versions as part of an anarchist faction). He took part in the battles with the White Army of General Deniken as it drove the Red Army towards Moscow and near-defeat. The tide was turned when Leon Trotsky, head of the Red Army, formed an alliance with Ukrainian anarchists and drove the Whites back, defeating them decisively in October 1919. By most accounts Timofèeff-Ressovsky was involved in the

# EVOLUTION AND EXILE

final battles but it seems that he fell ill with typhoid and returned to Moscow.

Quite how much of his education he managed to complete is unclear but in 1922 he was invited by Kolt'sov to join his research group, where he worked on *Drosophila* – presumably descendants of the flies brought over by Muller. His career took a fateful turn in 1926 when Oskar Vogt was appointed to head the project to study the brain of Lenin – who had died in 1924. Vogt was intrigued by some of Timofèeff-Ressovsky's *Drosophila* work and invited him to where he was to set up a genetics research unit at the Kaiser Wilhelm Institute. He and his wife and collaborator, Helena Alexandrovna, moved to Berlin that year.

It was here that Muller met him in 1932 and here that he had the fruitful cooperation with Zimmer and Delbruck. He was asked and then, in 1937, told to return to Russia. He declined and continued to work in Berlin during the Second World War until it was taken by the Russians in 1945. He was arrested and sentenced to ten years in the Gulag. He served but two of these – enough time to nearly die of hunger, develop all kinds of unpleasant conditions and almost lose his sight – but was released in 1947 – it was said at the insistence of Frederick Joliot-Curie – to continue his scientific work. This he did in various senior capacities virtually to his death in 1981. The cloud that hung over him because he worked in Germany under the Nazis was not lifted until 1992 when he was completely rehabilitated.

## *GENETICS IN RUSSIA*

Three group were the centres of genetic research in the USSR after the revolution: two in Leningrad and one in Moscow. In Leningrad Vavilov's Institute of Applied Botany (later the Institute of Plant Industry) was known for plant breeding – it had possibly the largest collection of seeds in the world – while Yuri Filipchenko's Department of Genetics and Experimental Zoology at the University. The two groups were combined under Vavilov's leadership when Filipchenko died in 1930 and some of the genetic activity moved to Moscow in 1933 to form the Institute of Genetics. The Moscow group at Kol'tsov's Institute of Experimental Biology (which opened in 1917) was made up of a section led by S S Chetverikov, with interests in the role of genetics in evolution, and another by A S Serebrovsky with more general genetic interests. *The Mechanism of Mendelian Inheritance* (Morgan et al's book) arrived there in 1919 and *Drosophila* work started soon after. However, it seems to have been Muller's visit in August 1922 with his cultures of *Drosophila* mutants and his accounts of what was going on in the USA that really triggered interest.

# GENES, FLIES, BOMBS...

Chetverikov started pioneering work on the genetic variability of free-breeding *Drosophila* populations (he was a lepidopterist by training) and, in 1926, he published the milestone paper on the mechanics of evolution that resolved many of the issues seen to exist between natural selection and Mendelism. Chetverikov himself was arrested in 1929 and exiled. His group dispersed.

Alexander S Serebrovsky made his name as a chicken geneticist and breeder particularly studying the geographic distribution of the genetic variation of the birds. It is said that he chose genetics because of lack of good equipment such as microscopes and the cheapness and simplicity of work with *Drosophila*. Originally a supporter of the presence-absence theory, he was persuaded (by his own *Drosophila* work with his pupil Dubinin that indicated intragenic structure) that it was wrong. His thoughts on population genetics – that derived from his field studies – anticipated later ideas about genetic drift and the role of population isolation in evolution. A supporter of eugenics, in 1929 he proposed a large-scale human artificial insemination programme – just before eugenics went out of fashion in the USSR.[22]

## *MULLER IN RUSSIA*

Muller joined the Institute, where there was already *Drosophila* work in progress, and began to build up his team, inviting Carlos Offermann and Daniel Raffel to join him.

He returned from a conference at St Moritz in July 1934 to find that Jessie had left him and taken David back to Austin. Attempts at reconciliation failed and they were divorced in May of the following year. David would stay in the Texas and Muller would have access for three months a year. Jessie was soon to marry Offermann.

In December 1934 he moved to Moscow where he seemed to enjoy the social scene with nights at the ballet and parties in his modest apartment. for American visitors. He employed Jenny Levy – who had worked for the English-language Moscow Daily News – as his secretary. He, perhaps through her, became acquainted with the colony of émigrés that clustered around the News and included Robert Talbott Miller, Jenny's future husband, writer Anna Louise Strong and journalist Milly Bennett.[23]

His work over the three and a half years he was in the USSR was perhaps not by his own standards exceptional. But, with his co-workers, he established three important principles:

> some genes could be shifted, even to another chromosome, and still function. While the function of some genes could be affected if they

were moved towards a heterochromatin region in the position effect – and would regain their function if moved back

the Bar mutation is a duplication suggesting the idea that new genes arise, in evolution, through duplication and subsequent mutation

genes are separated by inactive regions of the chromosome.

### IN MEMORY OF V I LENIN

He contributed to a book *In Memory of V I Lenin*, published in Russian and English[24] in 1934. The three striking claims that emerge from his polemic praising the 10-year-dead leader are that: genetics would have developed much more directly and quickly if Lenin's ideas had been followed; Thomas Morgan stole the ideas of his students and bourgeois, capitalist eugenics were bad but socialist eugenics good.

Sadly the precise ideas that would have helped so much are not specified or described. Indeed it would have been necessary to "realize and correctly to apply the general principles advocated by Lenin with regard to the basis and the development of scientific thought". It seems that this would have avoided the deviations, "unencumbered with the psychological impediments of the past age". Lenin's enemy in science was idealism; he was a materialist. Muller saw idealism in the shadows: in those who had resolutely opposed the chromosome as the source of inheritance; those who saw the gene as just a concept, the biometricians in general, Bateson and Karl Pearson in particular (Lenin had himself attacked Pearson). These people were not just wrong but victims of "feudal and decadent bourgeois attitudes" who fabricated "inconsistent muddles". How ironic that, once he found his feet in the USSR, Muller was to fight bravely with state-sponsored Lysenkoism, one the greatest corruptions of science in history.

In a rather sour passage, not really appropriate for a eulogy, Muller complained that it was "students" who had planned and interpreted key experiments and so forced Morgan to accept the chromosome theory. The younger *Drosophila* workers "had working-class connections, were class-conscious, and had absorbed a Marxian, materialist viewpoint".

Muller closes with a dismissal of social Darwinism and capitalist eugenics with their simplifications and perversions, their encumbrances of religious superstitions and customs, as key tools of extreme fascism. On the other hand, he and Lenin both recognised that there were marked genetic differences across nations. It was a socialist duty "to allow the individuals of later generations to receive as advantageous as possible a genetic equipment". They might then "enjoy increasingly that world conquest, on the path of which Lenin helped so much to set us." Not everyone was to appreciate the distinction Muller drew and that was to end, a few years later, his stay in the USSR.

Wrapped up in a vocabulary and intellectual framework alien, after nearly a century, to modern, more pragmatic minds, the piece is nonetheless an argument for an attitude – called here materialism – that is central to modern science. With the spice of bitchiness towards Morgan and the ambiguity about eugenics the article it is more interesting than its official title, *Lenin's Doctrines in Relation to Genetics*, promises.

## LYSENKO

Muller left Russia in 1937, dismayed by the growing influence of Trofim Denisovich Lysenko and the dominance of Michurinism, a set of ideas promoted, hijacked and expanded by Lysenko and his associates.

I V Michurin had been trying to improve fruit trees since he set up an experimental orchard at Tambov, about 300 miles south east of Moscow, in 1875[25]. He tried acclimatisation: bringing plants that grew readily in the south and trying to acclimatise them to the more northern climate. The procedure failed as did grafts of southern varieties onto local stock. He tried mass selection and found that it worked in some degree but that there was really little improvement occurring. The breakthrough came just before the Revolution when he began hybridising varieties and species from widely different habitats. He started a programme of collecting plants from across the USSR and the world and he claimed that by 1919 he had created 153 new varieties of fruits, nuts, tobacco plants.

Hybridisation was, of course, a widely practised technique and it was well known that vigorous new varieties could be produced by suitable crosses. To this extent Michurin's results were nothing new. However, he regarded grafting as a form of hybridisation (vegetative hybridisation) and believed that the characteristics of the grafted plant and the stock (or "mentor") were passed on to offspring. He also espoused the inheritance of acquired characteristics, believing that plants in a receptive state, perhaps the very young or after some shock, could be trained to become new varieties under the right growing conditions.

Whatever his methods, he claimed to have produced a large number of novel food plants. Lauded in communist Russia, particularly after his death in 1935, the reality and value of what he achieved remains in dispute until today.

Michurinsk, a large town in Tambov Oblast, was named after him in 1932. A film "Michurin (Life in Bloom)" was made about his life in 1948.

## EVOLUTION AND EXILE

A Russian Luther Burbank, he would perhaps have been remembered as an unscientific and moderately successful plant breeder but for the activities of Trofim Lysenko, who more-or-less single-handedly invented Michurinism.

T D Lysenko worked at an agricultural research station in Azerbaijan. His claims of how to grow crops without fertilizer and the production of winter peas in Azerbaijan, brought some recognition in a 1927 Pravda article but it was his "discovery" of vernalisation of wheat in 1929 that brought him national fame.

Winter wheat is a type that can be sown in the late autumn to sprout and root before the frosts. It lies dormant through the cold weather of the winter and then comes to life and gives a crop in early summer. The winters of 1927-8 and 1928-9 were harsh with little protective snow cover in the Ukraine and the winter wheat harvest was a disaster. Lysenko persuaded his father – a peasant farmer – to soak winter wheat seed in water and bury it in snow and then sow it as spring wheat. The small sample, it was claimed, not only thrived but gave a greater grain production than conventional spring wheat, normally sown. It was a technique that had been known for over 70 years and had been studied experimentally since WW1. However, Lysenko coined the term "vernalisation"(jarovizacija) and by mid-1929 was something of a celebrity – someone who had solved the problem of the failure of winter wheat. That it was on the basis of just one peasant trial field didn't matter. Lysenko was transferred to Odessa research station, the most important one in the Ukraine.

The further trials were rather disappointing; the yields from them were patchy and fell below the original expectations. He was later to claim a more dramatic effect still: with the right degree of vernalisation over a few years, he could change winter wheat into spring wheat permanently.

In the following few years Lysenko expanded vernalisation to include other crops (notably cotton and potatoes) and built a "phasic" theory of plant development to support it. He was given a journal, the *Bulletin of Vernalisation* in 1932 which failed in 1933 but was revived in 1935 as *Vernalisation* and renamed *Agrobiology* in 1946 (to reflect Lysenko's much broader interests by then) .

Lysenko turned his attention to plant breeding after a government decree issued 3 August 1931 that plant breeding for new varieties should be dramatically speeded up so that the traditional peasant varieties could be replaced with higher yielding ones on the new collectivised farms. The demand was that for some key crops (notably wheat) the new varieties must be available within two years. Generally the breeding cycle for new varieties of crops was to be reduced from 10-12 years to 4-5.[26]

Others promised and prevaricated (they could do little else) but within two years Lysenko, making use of vernalisation in his breeding programme, could promise a new variety of spring wheat that had been developed in less than three years and would be available by 1935. In 1935 he announced in his new journal that four varieties had been created. The success was put down to his theory of plant development allowing him to predict the outcome of crosses before they were actually made and therefore to eliminate extensive and abortive experiments.

Lysenko also promoted vegetative hybridisation from around 1940. This was of course a key element of Michurin's work with fruit trees and Lysenko initiated work on tomatoes and other members of the *Solanaceae*, notably potatoes. It was claimed that a graft of one variety of tomato onto another (the stock) produced seeds that grew into plants with characteristics of both the stock and the graft and so was a true hybrid Lysenko seems to have kept a specimen of a "hybrid" on his desk to show to visitors. The result was never obtained by anyone in the West and the Russian results were believed to have come from accidental cross-pollination of plants, poor experimental methods generally, the appearance of chimeras (where tissue with different genetic origins grows side by side) or by somatic mutation. However, Lysenko was still promoting vegetative hybridisation at a conference in 1948.

Later, at its highpoint, followers of Lysenkoism (although the term at the time was Michurinism) claimed to have transformed wheat into rye, plant tissue into animal tissue and to have extracted chicken from a rabbit.

It would be reasonable to ask what the theory behind these dramatic assertions and claims was. Jarovsky[27] suggests that this is not an easy question and calls Lysenko's ideas "vague and personal dogmatism" and suggests they "resist coherent presentation." His approach "began and ended with opposition to clear-cut thought and rational experimentation."

In fact, it may help to start with what he rejected – which was essentially all aspects of the genetics that had been so methodically investigated and so painstakingly assembled into a theory that accommodated Mendel's experiments, the distinction between genotype and phenotype, the chromosome and its mapping. Lysenko rejected all of that. He and his followers spoke with contempt of the "Mendel-Morgan-Weismann" theory and called it mystic, metaphysical and pseudo-scientific. He dismissed statistics and good experimental design as wastes of time.

# EVOLUTION AND EXILE

What he had in its place seems now to hark back to earlier times. The hereditary of the organism came not from chromosomes and genes but from the whole organism. As organisms went through various phases of development their heredity might be "shattered" by some stress (as Michurin had claimed), making them more pliable and able to take a new form better adapted to the new environment. A quote about species change from Lysenko gives an indication of the scientific quality of the thinking:

> We conceive the matter as follows: In the body of a wheat plant organism, under the influence of appropriate conditions of life, granules [krupinki] of a rye body are born. But this birth does not arise by means of a transformation of the old into the new, in this case of wheat cells into rye cells, but by means of the rise in the depths [nedrakh] of the body of the organism of a given species out of substance [veshchestvo] that does not have cellular structure, of granules of the body of another species.[28]

Of course, valuable practical results can come without meticulous experimentation and sound theory; we might otherwise not have had the wheel, the steam engine and penicillin when we did. However, Lysenkoism produced only disappointment and disaster. Vernalisation did not fulfil its promise, vegetative hybridisation did not work at all, species transformation was a chimera and Lysenko's final fling of cluster planting of trees was a national disaster.

Stalin's Great Plan for the Transformation of Nature was to turn great tracts of Russia from arid cold wastes to moist and temperate gardens. It was to be done by having peasants plant vast numbers of trees as windbreaks and Lysenko proposed a particularly efficient way of doing this. The trees would be planted in small clusters which should, by more conventional wisdom, be thinned out after a time to give the healthier ones room to grow. Lysenko saw that this was wrong because he had come to believe that natural selection did not work within a species; different members of a species worked together for the good of the species as a whole. So, he thought, the trees should be left un-thinned, the weak saplings would sacrifice themselves and would even link their root systems to those of their more vigorous neighbours to help them survive and flourish. The planting programme started in 1948, with about half the trees being managed according to Lysenko's recipe, and within a few years most of Lysenko's trees had died and very few that survived were actually flourishing. The Great Plan was abandoned in the early 1950s.

From the early 1930s to the mid 1950s he was an important figure in Russian science; from the mid 1940s he completely dominated genetics

# GENES, FLIES, BOMBS...

Many of his critics were suppressed by the regime; some of them, like Vavilov, were incarcerated and died (Soyfer lists ten eminent biologists who were persecuted with 8 of them being killed or dying in prison; Jarovsky lists around 60 botanists and crop specialists who were "suppressed" in the 1930s but there were quite possibly thousands associated with biology or agriculture who were persecuted and killed[29]. There were three conferences called on genetics in 1936, 1939[30] and 1948. The earlier ones may have been attempts by the authorities to find some kind of compromise between the Lysenkoists and the formal geneticists (another term used for the Neo-Mendelists) but by 1948 dissent from Lysenkoist views was outlawed and Lysenko could announce at the conference that he had the support of the Central Committee of the Communist Party.

> In August 1948 a session of the Lenin Academy of Agricultural Sciences of the U.S..S.R. devoted to a discussion of the situation in biological science was held in Moscow. At this session the Weismann (Mendel-Morgan) trend in biology was completely exposed and ideologically routed, as an anti-scientific, reactionary, idealistic-metaphysical trend, divorced from life and sterile in practice, in contrast to the Michurin trend, which represents the creative development of Darwin's teaching, and is a new and higher stage in the development of materialistic biology. The keynote of this discussion was Michurin's famous motto: "We cannot wait for favours from Nature; we must wrest them from her."[31]

Lysenko achieved great eminence in the USSR: President of the Lenin All-Union Academy of Agricultural Science, Deputy Chairman of the Supreme Soviet Council, Director of the Academy of Science's Institute of Genetics and Member of the Soviet Academy of Sciences

However by the early 1950s there was a growing recognition by the authorities that his methods were not working – the failure of cluster planting being the latest example – and criticism of him was implicitly allowed after the death of Stalin in 1953. This grew over the next decade in spite of intermittent support from Stalin's successor, Krushchev, until Lysenko was officially discredited in 1964 after Krushchev was ousted. His downfall came after a claim that he could improve the quality of milk from Russian cows. He produced bulls from Jersey stock that were supposed to create offspring when mated with cows in collective farms which would go on to produce plentiful milk of very high butterfat content. This high output would, Lysenko claimed, be sustained through many generations, if not forever. A report to the USSR Academy of Sciences published in September 1964 found that average yield per cow on his farm had in fact dropped by around 30% and that the butterfat content of their milk dropped, as might have been expected, with the degree of kinship with the original Jersey bulls. So the end finally came for Lysenko. He died in Moscow in 1976 but it has been suggested that genetics in the USSR did not recover until the 1980s: "The nightmare of Soviet genetics was, even in 1980, not quite over."[32]

If the experiments were flawed (and sometimes fraudulent), the theory was opaque, atavistic and wrong, and the results of the application of his ideas disappointing or disastrous, just how did Lysenko come to succeed for so long? Many reasons have been suggested. To some historians he simply offered quick, practical solutions to problems faced by the centralised bureaucracy (after collectivisation for example) when more conventional scientists proposed just more experiments. His peasant credentials were a bonus which meant he fitted into programmes to de-intellectualise science and allowed him, in the beginning at least, to encourage the disenchanted peasantry to engage with the bureaucracy's schemes rather that sabotage them. Other historians emphasise the way Lysenko's theories suggested a malleability of nature that meant, with guidance from the Communist Party and Marxism, anything could be achieved and the USSR could become an agricultural paradise. The guidance came from I I Prezent, a Communist Party member and Marxist philosopher, who operated with Lysenko from 1930, acting as some kind of political advisor and ensuring that Michurinism had the right political slant. The virulent criticism of conventional genetics (called some combination of Mendel-Weismann-Morgan-ism) identified it as reactionary and a bourgeois, idealist creation of capitalism. Yet other historians find him an unscrupulous and determined charlatan with an astonishing ability to conceal his own repeated failures from a desperate blundering bureaucracy with a fabric of duplicity and pseudo-scientific twaddle. Most likely all three factors apply.[33]

## MULLER AND LYSENKO

In his speech to the 1936 genetics conference Muller very bluntly attacked the ideas of Lysenko, Prezent and what he called the Neo-Lamarckists as "irreconcilable with the fundamentals of genetics." He set out very plainly the evidence for the gene ("as thoroughly established a fact as that of the atom") from both breeding studies and the cytology (including the new giant salivary chromosome studies) and linkage studies that tied them to the chromosome. He attacked some of Lysenko's treasured ideas of the time, notably those on in-breeding and cross-breeding. The differences between Neo-Mendelian genetics and the version promoted by Lysenko was analogous to those between "medicine and shamanism, between astronomy and astrology, between chemistry and alchemy."

Lamarckism was erroneous and leading to the dangerous and vicious conclusion that the effects of a poor environment would be passed down to offspring and so persist and become permanent distinctions between races and the rich and poor.[34]

This was greeted with applause by the Neo-Mendelians present and abuse by the Lysenkoites. Muller was later ticked off privately by the two senior Party members organising the conference for promoting racism. One of them did not believe in genetic effects at all.

This face-to-face polemic could hardly be bettered but Muller continued long after he left Russia to be a bitter opponent of Lysenko. Twelve years later he would write that Lysenko's vernalisation gave him "no more claim to being a geneticist than does doctoring dogs for worms."[35]

## *MULLER LEAVES USSR*

At the 1936 conference Muller put himself firmly on the side of real genetics and it became plain to him over the next few months that this was no place to be if you contemplated a future in the USSR. In April 1937 he tried to get published the bibliography of *Drosophila* genetics he had been compiling. The proofs were scrutinised by the Academy of Sciences and he was asked to omit reference to "the traitor" Agol and to the work of other arrested Russian geneticists, émigrés and German scientists. Reference to Agol was omitted but Muller managed to persuade the Academy to leave in all the others. The bibliography was anyway never published in Russia but did appear in 1939, published in Edinburgh while Muller was there. This was not just the influence of Lysenko but reflected something much deeper and more terrible. By now colleagues and friends were being harassed and arrested as part of Stalin's reign of terror.

Israel Agol's persecution had started even before he left Muller in the USA. Stalin had resolved the dispute between the dialecticians and the mechanicians with a 1931 decree ("Diamat") that said that dialectical materialism was limited to Marxism-Leninism, so supporting the mechanicians. It left Deborin and his followers seeming to close to Hegel and not close enough to Marx. Deborin managed to survive by keeping his head down but Agol found himself suspect and in 1936 he lost his job and was labelled a "menshevising idealist" and an enemy of the people. He was arrested and executed on 10 March 1937

Solomon Levit's mistake seems to have been to oppose Lysenko and to have supported – not least by his association with Muller – some version of eugenics. He had prospered somewhat after his return from Texas and become director of the Maxim Gorky Scientific Research Institute of Medical Genetics (MGI) but he disappeared in late 1936, not long after the conference where genetics had been described as a "fascist pseudoscience" and it was claimed by the Lysenkoists that he had been "unmasked". He was officially removed as Director of the MGI in July 1937. In January 1938 he was arrested and taken to the secret police's infamous Lubyanka gaol in Moscow. Here he was accused of being a Trotskyist, a traitor and an American spy. He was executed there, probably in May 1938.

It is uncertain how many other geneticists suffered for their science at around this time but Joravsky[36] estimated that 22 well-established workers in the field were "suppressed" – meaning anything from being unable to work for many years to being killed. He estimated that some 55 other biologists and agricultural specialists suffered similar fates in the Lysenko era.

Vavilov began to come under attack too, as we will see later, but there was something more personal, directed at Muller himself.

Muller had completed his eugenic work, *Out of the Night* (of which, more later), in 1934 and it was published by Vanguard Press in 1935. The ideas in it – such as artificial insemination – fitted in with rational approach to relationships and sex that Muller supposed prevailed in the USSR. With the encouragement of Levit – an influential Party member at the time – he sent a copy directly to Stalin in May 1936 with a covering letter suggesting a USSR-wide eugenics programme[37]. The work was translated and Stalin started to read it in early 1937. Soon after Muller got word that Stalin was "displeased by it" and had "ordered an attack prepared against it."[38] It would not be reviewed.[39] Muller had, against advice from friends, provoked the disfavour of a murderous tyrant – never a good career move – and it was time to be elsewhere.

Leaving was not without its perils for him and for those who might be seen to have helped him leave. With Vavilov he therefore worked out a plan to go first to Spain as part of the International Brigade and support the Republican side in the Civil War that had begun in July 1936. He would work with the Canadian surgeon Norman Bethune who was setting up a mobile blood transfusion unit. After his arrival in Madrid he worked on extracting blood to use in transfusions from fresh corpses[40] for just six weeks then, disillusioned with the administrative chaos that surrounded the unit, left in May 1937.

## *MILLY BENNETT*

The time in Spain may not have all been devoted to the gruesome. He apparently found time and energy for the continuation of an affair, started in Moscow, with Mildred Jacqueline Bremler – better known as Milly Bennett.

Milly, born in San Francisco in 1900, was a journalist in China in the mid-1920s, reporting on the struggles of Chiang Kai-shek's Kuomintang and then in Russia where she wrote for and edited the English-language Moscow Daily News between 1931 and 1936. She married a young gay Russian ballet dancer, Evgeni Vasilivich Konstantinov, in 1932 (apparently to give him a heterosexual cover) but he was arrested on moral charges in 1934 and sent to Siberia. In 1936 she arranged to be sent to Spain to be with Wallace Burton, the twin brother of a former lover, Wilbur Burton. Wallace was killed in action soon after she arrived. She worked for the Press and Propaganda Service of the Republican Popular Front and reported for AP, UP and the London Times.[41]

One of her friends in Moscow described her enormous appeal to men, explaining her long string of lovers:

> Milly was a homely woman, but she was blessed with an extraordinary figure. She didn't dress in a particularly sexy way, preferring the business skirts and blouses of the rather scruffy newspaper business. But her shapely figure turned the head of many a man with a roving eye. She was thirty-nine but looked years younger. Her face reflected her travels, her features craggy and rough-hewn. She was regarded as "one of the boys" in the newspaper office and at the café bars where the journalists, [a] crowd that included few women, gathered.[42]

The same friend asked a journalist what her appeal to men was and received the tantalizing reply:

> "Have you ever danced with her?" the newspaperman asked. "No, of course you haven't," he added with a wink that suggested Milly's charm lay not just in her ability to gather and write the news.[43]

We know little of the relationship between Milly and Muller[44] but it must have enlivened those grim weeks drawing blood from cadavers. After a visit to the USA to see his son and some time in Europe, he returned to the USSR. Here his, albeit brief, participation in the Spanish Civil War had restored his standing enough for him to be allowed to leave without incident and without damage to his Soviet friends. This he did on 23 September 1937.[45]

## NIKOLAI VAVILOV

Nikolai Ivanovich Vavilov was born in 1887 and educated at Moscow

Agricultural Institute from which he graduated in 1910 with a dissertation on snails as pests. He spent some time in Europe in 1913-14, working with Bateson at the John Innes Institute. On his return to Russia he rather quickly climbed the professional ladder and by 1922 he was heading the Institute of Experimental Agronomy which became the V.I. Lenin All-Union Academy of Agriculture in 1930; from 1930 to 1935 he was its first president. From 1930 to 1940 he was also director of the Institute of Genetics in Moscow. He was awarded the Lenin Prize 1926 and was honoured throughout his career in numerous other ways.

His greatest interest was in improving crops and to this end he travelled the world from the Americas to China collecting around a quarter of a million seeds and plant specimens that were kept at his seed bank in Leningrad. His most notable administrative achievement was setting up some 400 breeding research stations around the USSR but his name lives on in botany and plant breeding textbooks through two laws he formulated in the 1920s. His Law of Homologous Series and Hereditary Variation which stated that the more closely related two species are the more resemblance there will be in the variations they show. The law was helpful in guiding plant breeders in what might be possible with a particular species drawing on what had been seen in their cousins.

He had, through his collection work, a great interest in the origins of food crops and his second law (usually termed a theory) said that the region of greatest diversity within species is most likely its place of origin.

His thinking had an influence on evolutionary biology and he was throughout a convinced Mendelian. It was this latter that made him a centre of opposition to the ideas of Lysenko and brought his career to a sudden and tragic end in 1940 when he was arrested. Charged with wrecking Soviet agriculture, being part of a rightist conspiracy and spying for Britain, he was sentenced to be executed in July 1941. This was commuted to life imprisonment but he died of starvation in a Saratov gaol in January 1943.

## *EDINBURGH (1937-1940)*

The Edinburgh Institute was established as the University's Animal Breeding Research Department under F A E Crew in 1919. Crew built it up with a Chair in Animal Genetics in 1928 (which he held until 1944) and a new building and new name (the Institute of Animal Genetics) in 1930. By then the department had 12 scientific staff and 13 visiting researchers. During his tenure Crew attracted some big names: Lancelot Hogben, Julian Huxley, Haldane and, of course, Muller.

Work was in progress at various times on sex determination,

reproductive physiology, breeding of farm animals as well as in more fundamental areas such as *Drosophila* genetics, cytology and, charmingly but instructively, the genetics of budgerigar colours, in which Crew himself was a considerable expert[46.] The *Drosophila* work had been initiated after the visit of Morgan and Sturtevant to the UK in 1922. Crew summed up the impact this visit had:

> In 1922 Morgan visited this country and attended the 11th meeting of the Society in June. The President of the Society, A. J. Balfour, at that time Lord President of the Council, was in the chair. With Morgan came A. H. Sturtevant. They arrived loaded with hat-boxes filled with *Drosophila* cultures and with lots of microscopic slides displaying the chromosomes. Morgan gave an account of the mutants of *Drosophila melanogaster* and Sturtevant compared genetically *D. melanogaster* and *D. simulans*. The enthralled audience had no difficulty in recognising how greatly *Drosophila* had contributed to the advancement of genetics, on account of its mutability, its reproductive habits, its suitability as a laboratory animal and its relatively simple chromosome constitution. To us it seemed that the vinegar fly was to be regarded as one of the most important immigrants to enter the United States – probably with bananas from South-East Asia. No one in the United Kingdom had been using an animal or plant that could possibly have yielded in so short a time the genetical and cytological information from which emerged the *Theory of the Gene*.[47]

Muller stopped off in Paris but in November went to the Institute. Here he had a a large laboratory and three research associates, notably Charlotte Auerbach. Auerbach's original intention was to carry on with her study of the nature of the gene using the development of *Drosophila* as a tool but Muller quickly convinced her that mutation studies would be more fruitful – changing the direction of her entire career. He seems to have had a similar near-hypnotic impact on all the *Drosophila* workers.

S P Ray-Chaudhuri was recently arrived from India and expecting to start a PhD on silkworms under Crew. However he was persuaded by Crew to work with Muller instead. The result was both a PhD and a paper that Muller was to quote in his Nobel address. It showed that the production of mutations by gamma and X-rays was independent of dose rate down to a low level suggesting strongly that individual mutations were the consequence of individual ionisation events – and that there was no threshold before genetic damage started.

One of the other research associates was Kenneth MacKenzie and Muller encouraged and helped him to study the mutagenic effects of ultraviolet radiation. Altenburg had found in 1930 that it caused mutations in *Drosophila*. Towards the end of the decade Hollaender at the University of Wisconsin found similar effects in fungi and Stadler had found it had a mutagenic effect on maize. It seemed that ulraviolet caused mutations without the gross chromosome rearrangements that were caused by ionising radiation. It was interesting because of the light it might throw on the nature of mutations and thus on the structure of the genetic material in the dispute with Goldschmidt. The work with MacKenzie quite quickly confirmed that ultraviolet light did cause mutations with no visible sign of chromosome damage.[48]

The peak mutational effect was found to be at a wavelength of 265 nm, matching very closely the known absorption spectrum of DNA (and not that of protein). It suggested that nucleic acids might be an important element in the genetic material. These had been mentioned as possible genetic material by Wilson in his book *The Cell*... as early as 1896, but had fallen out of favour when analysis in the 1920s suggested they had a simple repeating structure and so were unlikely to be the carriers of complex genetic information. In the initial analysis of the ulraviolet results none of the researchers were willing to draw the conclusion that DNA *was* the genetic material – these were difficult experiments to perform with a number of possible confounding factors. When MacKenzie and Muller continued their work they found that the peak efficiency was between 300 and 320 nm – close to the absorption peak as others had found. Clearly there was some link here with the genetic material but they they still thought this was some kind of protein.[49]

With Guido Pontecorvo he developed a technique that allowed crosses between two *Drosophila* species: *melanogaster* and *simulans*. These involved triploid *melanogaster* females being mated with heavily irradiated *simulans* males, a process that resulted in embryos with *melanogaster* chromosomes except for a fourth chromosome from *simulans*. Analysis of the results allowed them to argue that the hybrids that failed to survive did so because of specifically genetic differences rather than as a result of some vaguer cytological incompatibilities. They also showed that faulty chromosome formation was the source of all cell deaths leading to spontaneous abortion of fly embryos.

## THE GENE GROUP

In October 1938 he went to Spa in Belgium for a meeting of Timoféeff-Ressovsky's Gene Group. This was the third of a series of informal meetings of leading geneticist and physicists interested in the nature of the gene. The series was an extension of the meetings that had led up to the Three Man Paper in 1935 and centred on the issues raised by it. The

## GENES, FLIES, BOMBS...

first meeting, on mutations and target theory, had been held in Copenhagen in 1936 and Muller had attended. He had cautioned on making too much of the results of target theory and the gene dimensions it suggested. The second was in Klampenborg, near Copenhagen, in April 1938 and he missed this. At the Spa meeting he presented a paper on the position effect.

The series of conferences was funded by the Rockefeller Foundation and supported by its director for natural sciences Warren Weaver. Weaver wrote a report of the 1938 meetings and used the term "molecular biology" to describe the area of study. A meeting was planned for Edinburgh to coincide with the VII Congress on Genetics. It never took place.

Over the summer of 1938 Muller returned to the USA on the *Queen Mary* for a visit that lasted two months. He spent much of the time at Woods Hole where he could work[50] and meet old colleagues. David came from Texas and they spent time together walking and sailing. He put out feelers for jobs. Reed College in Oregon rebuffed him; the University of Maryland seemed ready to offer him a professorship but the teaching load would be high and the research opportunities rather meagre.

In 1939 he wrote a report on the work so far for the UK Medical Research Council[51]. It warned that there was no threshold for gene mutation and other chromosomal damage and radiologists should be more cautious about using radiation – x-rays or rays from radium. It brought objections from Sidney Russ, the respected Secretary of the MRC Radiology Committee, that Muller had been rather reckless[52] in extending work from flies to man without qualification. Muller replied pointing out that similar damage occurred in other organisms and that there was no good reason to believe that it would not occur in man too. He wrote to leading UK geneticists for support suggesting, rather waspishly, that "geneticists will not wish to see the principles of their own science casually brushed aside by others who assume greater authority in these matters."

Some 600 delegates assembled in Edinburgh for the VIIth International Congress of 1939 but it was rather spoiled by the announcement of the non-aggression pact between Russia and Germany on its first day, the 23 August. The Germans left in a hurry, as did other Europeans, and all the delegates saw it as a sign of imminent war. There were no Russians there – Vavilov's plan had been to hold it in 1937 in Moscow as a demonstration that proper Soviet genetics was not dead – but the Lysenko affair was in full swing. Vavilov had resigned as President and Crew took his place.

The Molotov-Ribbentrop pact– as the pact is sometimes known – was the penultimate step in the brutal minuet that led to world war. Hitler had

# EVOLUTION AND EXILE

annexed Austria in the *Anschluss* in March 1938 and and threatened a similar fate for part of Czechoslovakia. The Munich agreement of September that year – between Neville Chamberlain and Hitler – avoided an immediate conflict at the expense of ceding the German-speaking part of the country to Germany. But Hitler proved to be not a man of his word and he annexed the whole of the country in March 1939. In April Germany rejected the Non-aggression pact it had signed with Poland and the Anglo-German Naval Agreement (which limited the size of naval forces and had been honoured by Germany only in the breach for some time anyway). By mid-August the USSR, Britain and France were discussing what would be done in the event of a German invasion of Poland – and were stuck on the fact that Poland was objecting to Russian troops ever being allowed onto their soil. So, while the build up to war came over a relatively long period, the world (and this included Communists around the globe and some of Germany's allies) was stunned by the announcement of an agreement between the Soviets and the Nazis. On 25$^{th}$ August Britain and Poland hurriedly signed a defence pact.

The congress went ahead with an unknown number of absentees, sustained largely by those who could not get away. Stoically and rather bizarrely, the conference spent two Plenary Sessions discussing the organisation of future congresses, wisely deferring consideration of an invitation to Rome in 1942. However it did take place; papers from absent members were read into the record. There were important papers on ultraviolet light and mutations by Stadler and Hollaender. Muller gave a paper on structural change in chromosomes[53] and Ray-Chaudhuri delivered his paper on the proportionality of mutations induced to total dose.[54] The planned Gene Group meeting did not take place after the congress closed but there was time to draft and sign the *Geneticists' Manifesto* – discussed later in the chapter on eugenics. The proceedings of the Congress were published in 1941.[55]

On 1$^{st}$ September, German troops invaded Poland. A Secret Protocol to the pact shared Europe (splitting Poland) between the USSR and Germany so, on 17 September, Stalin sent the Red Army into Poland to claim his share. War was declared on Germany by Britain and France and major members of the British Empire on 3 September.

The passenger vessel SS *Athenia* left Glasgow on 1$^{st}$ September, called at Belfast and Liverpool and set out for Montreal on 2 September. The following day it was torpedoed by German submarine U-30 a few hundred miles west of Scotland. The ship was slow to sink so almost 1000 people were rescued but 128 died, almost half of them in the course of the rescue effort. The dead included 28 American citizens. Among these there two participants in the Congress, from the University of Wisconsin, who had joined the ship, conveniently, in Glasgow.

Although rumbling prospects of war in Europe and the worry that his position in Edinburgh was probably but a temporary one must have been unsettling, Muller found his work satisfying and his associates creative and welcoming. Towards the end of 1938 he met and fell for Thea Kantorowicz, one of the technicians at the Institute. She was the daughter of a German Jewish surgeon who had been arrested and put in a concentration camp. He was released after pleas from the International Red Cross and the family settled in Istanbul. Thea completed her medical studies there and got the job in Edinburgh through family connections. She and Muller married in May 1939.

As time wore on into 1940 war began to have its impact. There were German war planes sometimes above Edinburgh, though no bombs fell. But there were irritations. Thea, a German still, was classified as a friendly alien and could still drive; Muller, a neutral alien, was not allowed to. His travel was restricted. He became unwell with an old eye problem flaring up. Pontecorvo was interned on the Isle of Man. Even work with the flies became restricted because of the blackout. In the summer he started to look for jobs in the USA. Although nothing turned up, they decided to leave anyway.

Leaving was not quite straightforward. Muller as an American was not able to travel on a British ship in wartime; Thea, with a German passport, might be restricted in travel in Europe. The plan that emerged was to take a ship to Lisbon and then fly on the Boeing 314 Clipper sea-plane[56] to New York. In the event the ship could not sail and they took a plane to Lisbon – accompanied by a selection of fly cultures in a bread box. It was now late August and they had to wait in Lisbon until 3 September for the flight to New York. The flies needed regular attention including feeding and had to be repacked into their slightly unconventional travel case. After they had been on the plane for a while Muller realised he had not seen the bread box for some time. A frantic search of the plane and even a radio call back to Lisbon failed to locate it. Sometime after all hope had faded a steward appeared, carrying the box labelled "BREAD", and explained that it had somehow found its way into the obvious place: the pantry.[57]

### *GENETICS IN THE EARLY 1940s*

Provine has summarised[58] the centres and workers who had made significant contributions to genetics in the 1920s and 30s. By the end of that period there was no doubt that the chromosome was the seat of inheritance and that the chromosome theory was entirely consistent with

# EVOLUTION AND EXILE

Mendelism. The gene concept – a linear arrangement on the chromosome with each gene associated with one or more characters, sometimes determining a character, sometimes influencing it – accepted by all but a few. Goldschmidt[59], for example dismissed the whole notion of the gene. The chromosome as a whole had to be regarded as the unit of heredity – not the gene.

The evidence for the gene/chromosome theory had come almost entirely from *Drosophila* studies conducted in the USA – and the ones elsewhere could generally be traced back to Morgan's group. Maize genetics was the other major source and this was entirely conducted in the USA. Up to the 1940s many other species had been studied but seldom at a fundamental level. However, the extensive experimental results of breeding could be interpreted using the chromosome/gene hypothesis – although heterosis remained something of a puzzle..

The formal nature of much of the fundamental genetics was summed up by the mathematician Hilbert in 1930:

> The numbers [percentages of crossing over] are in accord with the linear Euclidean axioms of congruence and agree with the axioms concerning the geometrical concept "between." Thus the laws of heredity emerge as an application of the linear axiom of congruence, that is, of the elementary geometrical propositions concerning the displacement of line segments — so simply and precisely, and at the same time so wonderfully that no one could have imagined it in his boldest fantasy.[60]

It is all too easy to take the ideas of the gene and of gene mutations that we all have in our minds today and paste them into the discussions and thinking of the 1930s. In fact they were concepts that arose only after complex experiments and procedures on insects and plants, detailed statistical calculations. The experiments involved creatures that had been subjected to immense doses of radiation which caused visible damage to the chromosomes as well as the invisible sub-microscopic mutations that were the supposed fine tools used in chromosome mapping. Even if the genes were arranged in strings as mapping suggested, there were problems like the position effect that shouted that the genes interacted. And, of course there was no chemical model of how these genes might be constructed.

The key information about chromosome organisation came from deductions from crosses between mutants, the mapping technique devised by Sturtevant in 1911 that was essentially very simple in concept but which left the gene at least as remote as the atom – as the Hilbert quote above indicates. The rediscovery of the giant chromosomes in the salivary glands of *Drosophila* rather changed that. Now the gene could nearly be seen; an indistinct image maybe (and of an anomaly) in just one creature but the link promised progress.

## *DOUBTS ABOUT MUTATIONS*

One area which was unresolved was connected with the effects of radiation and whether it caused mutations in the sense of changes to individual genes (point mutations), rather than rearrangements of existing, unmodified genes. As we have seen, Muller could deploy arguments based on the effects of ultraviolet radiation to show that point mutations were probably created. But his conviction that point mutations did occur was probably based largely on arguments from evolution: evolution needed changes in genes not mere rearrangements so changes there must be. Radiation he thought must be more than merely a wrecking agent.

Two doubts had arisen in Muller's mind while he was in Russia and persisted in Edinburgh, making MacKenzie's work particularly important. First the evidence for the existence of genes as discrete parts of the chromosome each with a distinct role in heredity was under attack – not least by Goldschmidt. Second the evidence that radiation could actually disrupt individual genes, as point mutations, came to appear less convincing.

Muller and Bridges had, separately, identified the bar mutation as being in fact the result of a duplication – see earlier where this is mentioned – and Muller's (and others') studies showed that there was a wide spectrum of chromosomal changes – deletions, translocations – after irradiation. The position effect that Sturtevant had discovered through studying the bar mutation meant that such changes could cause effects that simulated those of point mutations even though individual genes were not changed.

Lewis Stadler, who had discovered the mutational effects of radiation just after Muller, was doubtful that effects ascribed to point mutations (what he called chemical changes) after irradiation were actually more than just the consequences of chromosome breakage and rearrangement – what he referred to as mechanical changes. These could even, he thought, simulate one of the effects supposed to distinguish point mutations, the reversal of mutations by further irradiation, and, through the position effect, the consequences of rearrangements could be indistinguishable from those from point mutations. It was known that the changes induced by radiation in maize were anyway very different from spontaneous mutations. So Stadler spent much effort in deciding how to separate induced from spontaneous mutations and concluded that, at least in plants, the mutations induced by ionizing radiation were not the same as natural mutations. His results, he said, did not support "the assumption that mutation in general is affected by radiation."[61]

He did extend his studies to ultraviolet radiation, where he found that the induced mutations were indistinguishable from spontaneous ones, but these were never fully published[62]. While these supported the notion that mutations were changes within genes (intra-genic or chemical) rather than just rearrangements, they did not modify Stadler's conclusions about the effects of x-rays.

Belgovsky working with Muller in Russia saw tiny structural changes produced by radiation in numbers that were proportional to dose. Was it possible that there was damage (translocation, duplications, inversions), at a level that could not be seen in the polytene chromosomes, that caused effects that looked like mutations? It could mean that what Muller thought were mutations resulting from changes in individual genes were just the result of moving existing genes elsewhere. In his paper to the VIIth Genetic Congress he was not sure[63]:

> The problems are discussed of whether "gene mutations" are only ultra-minute linear rearrangements and of whether the "gene" is not a sharply defined segment of the chromonema. Evidence is presented showing that ultra-violet produces gene mutations without first breaking the chromosomes and this result as well as others is regarded as raising difficulties for such a view. It is held, however, that this matter is as yet far from settled, and the situation may be more complex than is generally realized, as indicated by studies on viruses.

Carlson has concluded that:

> The only basis for real gene mutations that Muller could rely on was his own conviction that evolution by gene mutation could not have resulted from a mere rearrangement of components already present in the chromosome. Clear proof of gene mutations defied genetic analysis until the late 1950's....[64]

## MULLER'S RATCHET

Evolution theory itself had incorporated Mendelian ideas and, with the notion of mutations that arose spontaneously to give ever-new material for natural selection and genetic drift to work on, it offered an explanation, if a developing one, of the process.

One aspect of evolution that continued to trouble theorists was the role of sex. Why was recombination so advantageous that almost all living things had adopted some form of it, at least sometimes. Fisher in 1930 and Muller in 1932 made an argument that was summarised by Crow as:

> ...a major advantage of sexual reproduction is that, by recombination, favorable mutations that occur in different individuals can be combined into the same lineage. Otherwise they can only compete with each other until one wins out.[65]

Muller estimated that sexual reproduction would increase the speed of evolutionary development by hundreds to millions of times over the asexual kind.[66]

When he returned to similar questions in 1964[67], Muller was to throw out a thought that was to lead to an understanding of the advantage of sexual reproduction. Talking about disadvantageous mutations he said:

> ...we find that an asexual population incorporates a kind of ratchet mechanism, such that it can never get to contain, in any of its lines, a load of mutations smaller than that already existing in its at present least-loaded lines.

The ratchet idea was first picked up by Feselstein[68] and he paid a tribute to the clear sightedness of Fisher and Muller:

> Until multiple-locus models involving genetic drift, selection, and recombination can be treated exactly or by diffusion approximation, quantitative conclusions on the effect of recombination may be difficult to draw. Given the difficulty of the problem, one could hardly have expected even FISHER or MULLER to have provided us with a quantitative theory of recombination. In retrospect, it is remarkable that they should have seen so much, so clearly, and so early.

Muller's Ratchet has been part of the increasingly complex mathematical modelling of evolutionary processes ever since.

## *MULLER AT AMHURST*

After their return to the USA Hermann and Thea stayed in New York hotel while Muller searched for a job. All the feelers he had put out himself came to naught; enquiries on his behalf by Huxley and Crow also drew a blank. His actions at Houston – those of a troublesome academic with radical views – meant that he could hardly expect a good reference from them; a term in the USSR was hardly a recommendation either. Step forward Harold Plough.

Like Muller, Plough had attended Morris High School but had then gone on to Amhurst College in Massachusetts, graduating in 1913. He had then returned to New York for his graduate studies at Columbia working on temperature effects in crossing over in *Drosophila*. He obtained his PhD in 1917 and, after a year on a fellowship, returned to Amhurst where he stayed for the rest of his career, retiring in 1959. Although Amhurst was, as it is still, an undergraduate-only college, Plough managed a productive research life. He published extensively, mainly on effects of temperature and (stimulated by Muller's discoveries) radiation on the fly's genetics.

He recommended Muller for a temporary research associate post at Amhurst and Muller cast of his residual links with Edinburgh and accepted. Amhurst, a liberal arts college about half way between New York and Boston was the alma mater of Calvin Coolidge, 30th President of the United States and the best known figure on the staff in Muller's time was probably the poet Robert Frost.

Muller joined the small biology faculty, chaired by the sympathetic figure of Otto Glaser, aware that he would not have the kind of facilities he had become used to elsewhere, certainly straight away. However, Plough and P T Ives had managed productive work on the fly, so he might too.

The research proved harder than he thought without assistants or graduate support and he actually, by his standards, achieved rather little. Apart from writing up the work he had done with Pontecorvo at Edinburgh, he produced just a handful of research papers, brief reviews and abstracts[69]. He did do some new work on *Drosophila* – studying spontaneous mutation in stored semen – but it was hardly groundbreaking. A study of the effects of the effects of ultraviolet radiation on chicken semen was inconclusive and never published. Just possibly, his most significant output was a well-received obituary of E B Wilson.[70]

He welcomed the proximity to Cold Spring Harbor in the summer and, in 1944, he and Thea had a baby, Helen Juliette, to enjoy but generally things were not going well. The wartime restrictions were irritating, the heavy load of undergraduate teaching was a burden and, perhaps worst of all, there was the nagging awareness that he was on a contract up for renewal annually. With staff cuts going on around him, it must hardly have been a surprise when he was told in early 1945 that his appointment would terminate in June. The long and disappointing business of finding something else, that he had started soon after his arrival at Amhurst, quite suddenly acquired great urgency. Now fifty-four years old and supporting a wife and new baby, with, seemingly, a tarnished past, he was on the verge of losing his job.

# GENES, FLIES, BOMBS...

## OFFERMANN AND RAFFEL

Offermann and Raffel's paths after Russia were very different but both seemed to find an entry into US academia difficult.

After their return from Moscow Offermann and Jessie moved to Chicago so that Carlos could finish the experimental work for his PhD. However in the following academic year she was diagnosed with TB and in the summer of 1940 they moved to the San Gabriel Valley in LA County in California in the hope that the condition would improve. This seems not to have happened and, instead, the condition steadily worsened. Jessie died in 1954. Carlos died in October 1983 in Monrovia, California.

Daniel Raffel was born in Baltimore on 7 August 1899. His father – Jacob M Raffel, a paper box manufacturer – had married Bertha Stein and they lived in west Baltimore with their children Daniel, Arthur S and Gertrude A. After High School, Daniel attended Baltimore Polytechnic Institute and was awarded a BA degree in 1917.[71]

When he registered for the draft in September 1918 he gave his occupation as a Clerk with J M Raffel & Co and whether or not he intended to go into the family business it was possibly forced upon him by the death of his father in 1919. In the 1920 Census he was down as a Paper Box Salesman.

In the early 1920s he met and married (in 1925) Rose M Mahr and must have started on a PhD at Johns Hopkins University. In 1928 he and Rose travelled to Europe and in 1929 he was awarded his PhD for his work with H S Jennings on the single-celled creature *Paramecium*. Soon after, he was awarded a prestigious National Research Council Fellowship and seems to have spent his time at Johns Hopkins, Yale University and the Woods Hole marine biological research centre in Massachusetts.

He was described by one of his colleagues at this time as "a very bright and forceful man". So forceful that, when he disagreed with Jennings over the interpretation of some experiments, he insisted that he should publish his work separately. Jennings, a much respected figure in scientific circles, told a colleague that "I cannot work with a man like that." and this view was shared by others in Jennings' group. It was agreed that relations between Raffel and the group would be severed – and it was at this point that Raffel was invited to work with Muller in Russia. Muller had gone to Germany for a time and then to Communist Russia after losing his job at the University of Texas because of his involvement with a Communist student newspaper – *The Spark*.

Raffel and Rose travelled to Russia in 1932 and stayed there until 1937

# EVOLUTION AND EXILE

working in both Leningrad and Moscow on *Drosophila*. The work was important and it threw more light on difficult aspects of genetics. They seemed to enjoy Muller's company with parties and visits to the Bolshoi ballet. They returned from Russia in July 1937.

Daniel seems to have tried to find a university post after his return but nothing was forthcoming. Possible employers were reluctant given the time spent in Russia and the association with Muller.

Quite what he did is not recorded until in round about 1942 when he took up farming and particularly cattle breeding, specialising in Ayrshires. When this became too hard physically, he took the post at Park School and taught there until his death in 1965.

There was another fascinating aspect to his life. His mother Bertha was a sister of the author Gertrude Stein and Daniel and Rose visited her in Paris on their trip to Europe in 1928. Gertrude wrote an enigmatic portrait of him: "Dan Raffel a Nephew". In 1946 he and Rose visited Gertrude again and, with her companion Alice B Toklas, was there when she died on 27 July after an operation.[72]

Muller met the Raffels when he arrived in New York (they had a house in Falmouth at the time) after returning from Russia and he stayed in touch with them after that. He visited the farm in Sparks in 1946 just after he learned he had been awarded the Nobel Prize – he told them and they were among the very first to know.

## *MISCELLANEOUS*

By 1940 much had been realised in the search for and understanding of the hereditary material and there was much more to come. Radiation had been an indispensable tool – even if its actions in causing mutations were not fully understood. However, it was to slowly decline in importance as a probe as more controlled physical and chemical techniques developed in the coming decades. This watershed for genetics in the 1940 and 1950s coincided, of course, with dramatic changes in nuclear physics and its applications. So, it was the hazards of radiation and, particularly for the next 30 years, the genetic damage it might cause that were to be major preoccupations of many geneticists, not least for Muller himself.

# 6 BEFORE THE BOMB

## X-RAYS

Wilhelm Conrad Röntgen was a 50 year old professor at the University of Wurzburg in 1895. He was operating a Lenard tube, a discharge tube with a thin aluminium window that allowed cathode rays to escape to the outside. The tube glowed where the cathode rays struck its sides but he noticed, when he covered it with black card, something else was glowing too: a barium platino-cyanide screen lying nearby. He quickly realised that he was dealing not with cathode rays but – because of the penetrating power – with something new. He worked in secret for the next few weeks investigating the properties of these new rays with a fluorescent screen and photographic plates. He found that the rays originated from the glowing area of the discharge tube and that they travelled not only several metres through air but through solid objects as well. On his photographic plates he found he could record the shadows the rays made as the passed through his wife's hand. The bones were clearly visible. His discovery was very quickly announced and within a few months it generated a kind of fever.

Since discharge tubes were common enough in laboratories, all round the world people quickly started taking "skiagraphs" as they were first called. Many were for pure entertainment but their medical potential was soon realised. Tubes were made more efficient by having metal anodes inserted and many other improvements in technique were found but they remained difficult to use ("a glass bulb surrounded by profanity" as someone put it) because they depended on residual gas in the tube to produce the electrons. That is until a copious source of electrons, a heated wire, was introduced by W D Coolidge in 1913, exploiting technology developed by the emerging electric light industry. The old gas tubes persisted for a while but the Coolidge device heralded the beginning of modern radiography with its intense and controllable supply of x-rays.

## X-RAY THERAPY

The effects of radiation on the skin – hair loss and burns – suggested that it might have some value in treating skin conditions and other ones which were resistant to then-current treatments. After all, ultraviolet radiation was already in use for some of these. So as early as December 1896 Leopold Freund made the first considered use of x-rays as a therapeutic tool when he irradiated a large hairy birthmark (hirsuites) covering the entire back of a 4 year-old girl. The treatments were spread over more than 10 irradiations each lasting 2 hours and the growth was removed but the patient suffered episodes of serious ulceration over many years. However, when she was examined in 1971, aged 75, she was found to be well.

By the early 1900s radiotherapy was being used quite extensively and successfully (and perhaps often unwisely) for non-malignant skin conditions conditions such as tinea capitis, acne vulgaris, eczema, lupus, skin tuberculosis[1] as well as for skin, breast and other cancers. Many who did used the technique saw it as a form of cauterisation rather than anything more sophisticated but during the 1910s it became rather clearer that it was indeed skin conditions that might benefit most with the equipment and techniques then available.

### X-RAY HAZARDS

The absence of shielding around the early x-ray tubes resulted in injury to the operators. The effects seen in the first few years were reported as sunburn, dermatitis, sore eyes, swelling and hair loss. In the summer of 1896 Herbert Hawks was demonstrating x-rays in Bloomingdale Brothers' Store in New York and the symptoms he developed included all the above; doctors treated it as a case of parboiling.

For a while these effects could be ascribed to other sources than the x-rays themselves. People blamed the electric sparks in the high-voltage generator, ultraviolet radiation, chemicals used in developing plates, ozone generation in the skin and faulty technique. However within a few years it became obvious that x-rays were the culprit – and that the effects were more persistent and serious than had been suspected. By the end of the first decade of the new century it was found that the disfiguring and painful wounds could become cancerous. Fingers, hands and limbs were amputated in attempts to prevent the spread of disease but many practitioners were to die of cancer (usually of the skin) or other conditions, such as anaemia, caused by the radiation.

Since a common way of checking a set-up was for the operator to place his own hand between the tube and the fluorescent screen, the invention of the Chiroscope in 1903 must have made some difference. This was a skeleton hand with simulated flesh mounted behind a fluorescent screen. The Osteoscope was a similar device using a complete forearm[2].

### X-RAY PRECAUTIONS

Shielding of the tube was unusual before about 1908 but some

## GENES, FLIES, BOMBS...

practitioners were careful throughout. Francis H Williams of Boston can be seen with a protective box around the tube in a 1902 photograph[3] and he remarked later that he thought penetrating rays like x-rays must have "some effect upon the system" and took precautions accordingly. Williams's early caution came from his brother-in-law and collaborator, the remarkable William Rollins. Rollins, a Boston dentist, put a guinea pig in a Faraday chamber – a set of electrically-earthed boxes that excluded any electric fields – and exposed it to an x-ray source outside the box. The exposure lasted two hours per day and, after 11 days, the guinea pig died. A second died, after similar treatment, after 8 days. It was not possible that the death was due to electrical fields – these were frequently blamed for the effects of working with x-rays and it led Rollins to propose three precautions:

> physicians should wear glasses that keep out x-rays when using fluoroscopes
>
> x-ray tubes should be kept in shielded boxes with a small window to give a cone of radiation no larger than needed
>
> patients should be shielded except where necessary for examination or treatment.

Rollins was a man ahead of his time. His impact was limited in the USA because the growing x-ray community was not disposed to accept that the astonishing new rays might have a serious downside that might limit their spread. His results and proposals may, anyway, not have been widely known outside the Boston area and they certainly did not cross the Atlantic.

By perhaps 1910 the dangers of acute and disastrous tissue damage were widely recognised and there were some straightforward protection measures being adopted. The means of measuring larger doses were available and were used for control of patient exposures. Together these things could, if sensibly applied, reduce and perhaps eliminate the dreadful acute effects – and within a few years professional bodies would step in with recommendations on protection to do just this. However, many of the early workers were to die because of their injuries and even more were to suffer and die from unsuspected long-term effects that were a long way from being understood. In 1936 a memorial was erected in Hamburg to the early pioneers of x-rays who suffered radiation injury or lost their lives due to their work. Of the original 169 names from 15 nations, 14 are from Britain.

While the association of skin cancer with radiation was clear within a decade of Röntgen's discovery, there was a perception that it was closely associated with the more obvious damage. If the obvious damage could be avoided then skin cancer would not follow.

For much of the first half of the century the possibility that radiation caused leukaemia was either disregarded or was controversial[4]. While a few cases were reported before the 1930s, it was only in 1931 that Aubertin reported five cases of myeloid leukaemia in radiologists when he had seen only one in other medical practitioners. Even after this the link was not regarded as conclusively established and when Colwell and Russ reviewed radiation injuries in 1934[5] (the year Marie Curie died) they could not decide whether leukaemia was one.

## RADIOACTIVITY

Henri Becquerel knew of Rontgen's discovery from a meeting of theFrench Academie des Sciences in January 1896 and thought that the x-rays might be connected with fluorecence in the tube – they seemed to originate from it.

He soon retrieved some crystals of uranium salts he had lent to someone else: they were phosphorescent when exposed to sunlight and could therefore be expected, if he was right, to give off x-rays. To test this Becquerel wrapped a photographic plate in black paper, put a coin on the top and then the uranium crystals on top of the coin. This package was then exposed to sunlight for several hours and, when the plate was removed and processed, Becquerel found a faint image of the outline of the crystal and the coin. This seemed to show that the luminescence of the uranium crystals, produced when they were exposed to the sun, had created x-rays. These had penetrated the black paper and caused the image on the film. He reported these results to the Academie on 24 February 1896.

He repeated the experiment a few days later but there was not much sun so he put the experiment in a dark drawer for several days. When he processed the plate, expecting nothing because the crystals had not been exposed to light, he found a very clear image. He initially thought there was some kind of stored phosphorescence but over the next few weeks he established that the radiation that exposed the plate came equally from other uranium compounds that were not luminescent. He also found that the radiation discharged electrified bodies – just as x-rays did. Since uranium metal gave very strong images he concluded that the radiation ("une phosphorescence invisible") originated from the element itself. He reported this to the Academie on 18 May 1896. There was hardly any public reaction to this discovery and and little scientific interest in it.

# GENES, FLIES, BOMBS...

Becquerel turned to other things but at the end of 1897 a young Polish woman living in Paris with her husband and daughter was looking for a subject for her doctorate and thought Becquerel's work might provide one. She began a systematic investigation of all the known elements to see if uranium was unique and discovered that there was one other element, thorium, with the same property. She coined the term "radioactivity" to describe it.

She made the dramatic discovery that some of the unrefined mineral samples were more active than could be explained by their uranium or thorium content. The ore pitchblende was nearly four times more active than the uranium oxide extracted from it. Her earlier work and this discovery, all resulting from the most meticulous work, convinced her that she was dealing with a new element. This was announced, as a probability, on 12 April 1898. To convert the probability into a certainty she would need to isolate the element from the pitchblende and she knew that it could be there only in minute quantities. Working with her husband Pierre, by July they were able to announce the discovery of the element polonium, a radioactive metal with similar chemical properties to bismuth. After several months further work they (with their collaborator G Bemont) announced a second element, radium, which was nearly 1000 times more radioactive than pure uranium.

## RADIUM THERAPY

The rays from radium proved to be not as useful for diagnostic purposes as those generated by x-ray tubes but what Marie Curie had discovered was an intense source of radiation that needed no source of power, as x-ray machines did, and promised great things in radiation therapy.

Industrial scale isolation of the element began in Paris in 1902 an by 1914 around 12 gm of radium had been separated.

The success of the Paris work led to the setting up of the Laboratoire Biologique de Radium there in July 1906 and within three years some 900 patients had been treated. Louis Wickham and Paul Degrais published their 1909 book on therapy *Radiumthérapie* with an English translation *Radiumtherapy* following in 1910. The book and the subsequent Paris work established radium therapy as an effective treatment and systematised the techniques involved. By the the outbreak of the First World War, institutes had been – or were being – set up in France, the USA, the UK and Germany.

## RADIUM HAZARDS

Although both Pierre Curie and Becquerel had experienced burns when

# BEFORE THE BOMB

they kept samples of radium about their person, there are few accounts of permanent damage caused by radium in the first two decades after its discovery. The material was quickly recognised to be valuable so that it was stored carefully and the inverse square law must have given a high degree of protection. It was also not as widely available as x-rays. .

The effects were unpleasant but not, compared with what had been seen for x-rays, very serious and were generally limited to the hands of people who had handled the element. Numbness and increased heat-sensitivity was followed by thickening of the skin, a loss of its elasticity and cracking of the nails.

The experience of George S Willis, an early entrepreneur in the US radium industry, can stand as an example. Willis worked with x-rays from 1905 to 1917 and from 1912 he also handled radium salts in glass tubes "freely and with no attempt at protection" several times a day: the amounts were significant fractions of a gram. In 1918 he noticed a numbness in his fingertips and a weakness in his left arm. A tenderness and then a soreness developed in the fingers and then fissures, one so painful that it had to be treated – in a bizarre hair-of-the-dog antidote – with radium. The fissures worsened and ulcerated and when one of them was excised malignant tissue was found. There were other similar cases around the world.

There was also plenty of evidence of how dangerous radium could be from patients: while there were remarkable cures or remissions there were also some terrible injuries from ignorance, carelessness and misjudgement. They could be regarded as normal risks associated with the necessary large doses. For workers, for a long time, the effects seemed to be limited to the hands. However, by the early 1930s evidence of the more sinister consequences began to gather from radium as well as x-rays – which were now recognised as being similar in their effects.

In 1933 Colwell and Russ could point to 16 cases of leukaemia among x-ray and radium workers, with just two of them in radium workers. They recognised that this total was probably higher than it should have been but concluded that the question of whether radiation actually caused them was "sub judice".

## *RADIUM PRECAUTIONS*

The protection measures used in those early years of radium are not as well documented as those for x-rays. That they were basic we can surmise from some of the recommendations on protection against radium published by the X-ray and Radium Protection Committee in 1921 :

# GENES, FLIES, BOMBS...

In order to avoid injury to the fingers the radium, whether in the form of applicators of radium salt or in the form of emanation tubes, should be always manipulated with forceps or similar instruments and it should be carried from place to place in long-handled boxes lined on all sides with 1 cm. of lead.

In order to avoid penetrating rays of radium all manipulations should be carried out as rapidly as possible and the operator should not remain in the vicinity of radium for longer than is necessary.

## *PAINT AND RADON*

A difference between x-rays and radium was that the radioactive element, if it entered the body, could irradiate some body tissues for a very long time because it became incorporated into them. The consequences of this intimate association between radioactivity and living tissue proved disastrous.

The use of luminous paint - zinc sulphide with added radium or mesothorium (Ra-228) – seems to have begun in Germany in the 1900s and by the beginning of the First World War it was taken up in America. The US Radium Corporation became a leading supplier of luminous watches to the United States military during the war and it employed several hundred women at its Orange, New Jersey factory in painting the dials. In spite of an awareness by the company of the hazards of radium, the women were allowed (or was it encouraged?) to get their brushes to a fine point by licking them; each time they did this they ingested a small amount of radium. Grace Fryer, who had worked at the factory for just three years before leaving in 1920, had developed serious bone decay in her jaw by 1922. When an investigation was launched in 1925 by the New Jersey Consumers' League, other women working at the factory were found to have suffered similar effects. Dr E L Hoffman, who conducted the inquiry, found that 12 women had suffered from persistent infections of the jaw with, sometimes, anaemia; four had died as a result. It was also found that the company had been warned the year before, after a secret investigation of their own, of the dangers of the procedure and of the widespread contamination in the factory. Fryer and four other women started a lawsuit against US Radium in 1927 and, with the support of a press campaign and in spite of their declining health, won compensation in 1928.

The scientific work of Harrison S Martland, a county medical examiner, documented in some detail the medical effects and helped make the link with radiation. His publications between 1925 and 1931[6] established the presence of radon, the radioactive gas generated as radium decays, in the

# BEFORE THE BOMB

breath of the affected women (a clear indication of ingested radium) and established a reasonable epidemiological link with radiation. He found both the acute effects – as Fryer had experienced – and a much delayed condition; both led to necrosis and, in many cases, bone cancer. In the work reported in 1931 he examined 18 deaths from occupational disease among 800 young women all of whom had worked in the factory for less than two years. Most had severe anaemia but in five cases death was due to bone cancer. So, over a quarter of the deaths were from bone cancer and this was so different from what he expected (0.1% was the equivalent fraction he deduced from autopsies) that he concluded radioactivity was their cause. Martland's work continued to the 1950s as he followed up the cases of chronic poisoning.

Another even more sinister effect of radium emerged in the mining pitchblende, a uranium ore.

The mines at Joachimsthal (now Jachymov) and Schneeberg (now on either side of the Czech Republic/German border) were a major source of silver in the 16th century and, when this was depleted and became available from the New World, of bismuth and cobalt. The discovery of uranium in the pitchblende in 1789, led to renewed life to exploit the remarkable properties of uranium salts in porcelain glazes. The extraction became more difficult and by the end of the 19th century the mines were near to closure when Mme Curie discovered radium. So valuable did the ore then become that the mines were reprieved yet again; the last was closed in the 1960s.

The miners had suffered a high mortality from a chest disease known as "mountain sickness" for a long time and it was shown in 1879, by Harting and Hesse to be a form of cancer. It was Arnstein, in 1913, who concluded it was a lung cancer and he found that it accounted for nearly half the deaths of miners since 1875.

The association with radon was suggested in 1921 by a Schneeberg native, Margarete Uhlig and after measurements of the radioactivity in the air in the mine by Rajewsky in 1939 this was widely accepted[7].

## *THRESHOLD HYPOTHESIS*

The concept that dominated for these effects on the irradiated individual, the so-called somatic effects, was that they would only occur if the radiation dose was above a certain level. In the early days the notion was that, provided there was no obvious physical damage, there would be no adverse effects. In particular, cancer would be avoided if this threshold was not exceeded. Ideas about where the threshold might be changed with time. As time went on obvious physical damage was replaced by more transient effects, notably the erythema that resulted from large exposures. For a long time it was reckoned that provided the erythema dose was avoided (or some fraction of it) there would be no harm. The threshold notion persisted up to and through the Second World War – and well beyond – for these somatic effects.

The situation was, due to Muller, very different for possible genetic damage.

## MECHANISMS OF DAMAGE

After Roentgen's discovery scientists were not slow to study the effects of radiation on living things. While many experiments were unstructured and looked at the effects on the tissues, there were scientists looking at effects at the cellular level and in the cell nucleus. For example, in 1903 Bohn was able to conclude that the main effect of radium treatment on sea urchins was damage to the chromosomes. In the following year, after studying the effect of radiation on developing eggs of the parasite of horses *Ascaris*, Perthes suggested that the chromosomes of developing eggs were fragmented. This was confirmed in 1905 by Koernicke who treated *Lilium* (lilies) with radium.[8]

So, the fact that radiation could damage chromosomes and therefore disrupt subsequent cell division was known by the end of the first decade of the century. It was clear that rapidly dividing cells were more susceptible[9] and it seemed likely that the chromosomal damage was the cause of the somatic radiation effects that were by then known in people. It was Bardeen, in 1906[10], who was the first to show that irradiated toad sperm led to fertilized eggs that, after a seemingly normal start, failed to develop properly. Since the spermatozoa are entirely nuclear material this showed that it was the nuclear material – the chromosomes – damaged by the radiation that led to the deleterious effects. The results were quickly confirmed with others species such as frogs and rabbits[11].

However, it was the observations of Mavor in the early 1920s that showed convincing evidence of radiation damage to chromosomes. In 1921 and 1922 he showed that doses of x-rays increased the frequency with which chromosomes failed to separate in mitosis and in 1923 he showed that there was also an effect on the frequency of crossing over.[12]

Now while this meant that the damage to the chromosomes did affect the individual irradiated (and probably produce the obvious somatic effects) and might persist through a number of normal cell divisions, it did not show a true genetic effect. While the effects of radiation could be seen microscopically on chromosomes as breakages and other gross (in chromosomal terms) damage, there might be other more subtle changes that affected individual genes and could be passed on to offspring. The distinction between chromosomal damage and mutations of individual genes was to be an important, although occasionally fuzzy, one for the rest of the century and the understanding of the ability of radiation to cause mutations made its first major leap forward in 1926 with the work of Muller.

## MULLER AND HAZARDS OF RADIATION

Muller was not slow to point out hazards.

> The transmuting action of X--rays on the genes is not confined to the sperm cells, for treatment of the unfertilized females causes mutations about as readily as treatment of the males. The effect is produced both on oocytes and early oogonia. It should be noted especially that, as in mammals, X-rays (in the doses used) cause a period of extreme infertility, which commences soon after treatment and later is partially recovered from. It can be stated positively that the return of fertility does not mean that the new crop of eggs is unaffected, for these, like those mature eggs that managed to survive, were found in the present experiments to contain a high proportion of mutant genes (chiefly lethals, as usual). The practice, common in current X-ray therapy, of giving treatments that do not certainly result in permanent sterilization, has been defended chiefly on the ground of a purely theoretical conception that eggs produced after the return of fertility must necessarily represent "uninjured" tissue. As this presumption is hereby demonstrated to be faulty it would seem incumbent for medical practice to be modified accordingly, at least until genetically sound experimentation upon mammals can be shown to yield results of a decisively negative character. Such work upon mammals would involve a highly elaborate undertaking, as compared with the above experiments on flies.[13]

In an address to physicians in Waco Texas in 1928 he expressed his concerns about both the diagnostic and therapeutic uses of x-rays. In diagnostic use there should be training of personnel and irregular inspection of machines to ensure that the lowest possible dose was given. X-ray therapy should be restricted to those conditions where there was no effective and safe alternative. For both uses he reminded his audience of the need for shielding, for themselves and the patients, to avoid unnecessary exposure. He also, in an indication of his eugenic preoccupations, warned against using radiation as a tool to produce human mutant geniuses. The harmful effects would be far too great to justify this. The talk was not received too well; several of the audience stalked out and a proposed talk to a group of radiologists later in the year was cancelled.

A key question posed by Muller's work was the relation between the radiation dose given to the flies and the number of mutations that resulted. Hanson and Heys, by the end of the 1920s, showed that there was a proportionality between dose and mutations in *Drosophila* and that this was true for betas, gammas and x-rays[14]. Oliver, in 1930, demonstrated this over a 16-fold range and Muller and Ray-Chaudhuri removed any doubt with their work in Edinburgh.

So by the beginning of WW2 the generally accepted views about the effects of radiation on the individual exposed and on his descendants were at odds. It was widely thought that there would be no serious long-term somatic effects provided doses were kept below some threshold level. However, the likelihood of hereditary effects was proportional to the dose with no threshold and all doses, however small, brought some risk to future generations.

The two views arose from quite different foundations. The views on somatic effects were driven almost entirely by human experience and were therefore derived from a relatively small and unreliable database (not least because of the inadequacy of dosimetry of the time) of people who had, on the whole, experienced large doses of radiation. The ideas on hereditary effects, on the other hand, were deduced from well-planned experiments under laboratory conditions but these were performed on a fly – an animal so remote from us that the relevance of the conclusions was, at least, arguable. Also, chance or not, the ideas came from two generally distinct groups of people with different interests: the threshold principle arose from those involved in the use of radiation who were committed to its benefits (and were therefore content that there appeared to be a way of eliminating its long-term hazards completely) while the proportionality principle came from academic geneticists.

## *THE SOMATIC MUTATION THEORY*

One idea that might have given (and maybe did give) the thresholdists pause for thought was that of Theodor Boveri in his 1914 book[15]. He proposed what came to be called the somatic mutation theory of cancer suggesting that cancers arose from a single cell as a result of a re-arrangement of its chromosomes. If Boveri was a little vague about what this re-arrangement might be, others were quick to identify it as a mutation. Probably too vague, contentious and unsupported[16], even when combined with the near certainty that radiation caused mutations in proportion to dose, to make the threshold idea seem somewhat optimistic, it did point the way to a deeper understanding of the hazards of radiation.

# 7 THE BOMB

## MANHATTAN PROJECT

The fact that uranium atoms could split was discovered by the German Otto Hahn, working in Berlin, and his long-time collaborator, the Austrian Jew Lise Meitner, who had already escaped to Sweden when Germany annexed Austria in 1938. Hahn found in late 1938 that when uranium was bombarded with neutrons a new element was produced. After secret consultation with Meitner and some very delicate chemistry back in Berlin he concluded that the mystery element was in fact barium – astonishing because barium has an atomic number of about half that of uranium. The results were published tentatively in January 1939 and it was not long before he had found strontium and yttrium, with similar atomic numbers as well.

The insight that the uranium nucleus had split came from Meitner and her nephew and fellow refugee Otto Frisch. It was not difficult, with what was already known about the nucleus, to work out that the split would be accompanied by a release of energy and that this, while minute for an individual atom, would be gigantic – far beyond that from chemical reactions – if even a small mass of uranium could be persuaded to split. The conclusions were quickly confirmed by scientists in Denmark, France and the USA and an exciting, even terrifying, possibility arose. If, as seemed likely, the uranium atom emitted neutrons when it split, these might go on and cause another atom to fission. And so on in a chain reaction that would release the promised vast amounts of energy. It might be possible to control the release and generate useful power; it might also be released explosively and make a powerful bomb.

Over the summer several scientists tried to work out whether a chain reaction could be achieved practically. The problem was that the number of neutrons released in fission was only just enough to sustain the chain reaction. If too many neutrons escaped from the material then the chain reaction would never get going; a large enough quantity of uranium in a compact form would be needed to keep the escaping fraction low. The conclusions of the scientists' calculation were not encouraging: it might work but as much as forty tonnes of uranium in the form of a sphere would be needed[1] to form the necessary critical mass. It did not seem a good basis for a practical weapon.

# THE BOMB

Prospects changed dramatically in February 1940 when Frisch, now in Birmingham, and fellow émigré Rudolf Peierls calculated the critical mass for one of uranium's isotopes, U-235. They concluded that less than a kilogram of U-235 would make a critical mass. The problem was that less than 1% of uranium was U-235. Since it was chemically identical to U-238, which made up the overwhelming proportion of the element, it would not be easy to extract the potent isotope. However, the Memorandum they wrote to Government started experimental work to confirm their calculations and prompted serious thinking about how U-235 might be separated. Frisch and Peierls were not part of this because they were aliens.

By the middle of 1941 it was clear that a uranium bomb was feasible and the explosive power, thought to be around that of 1800 tons of the chemical high explosive TNT, could make the weapon decisive in the war. It was decided to share the information with the USA.

The USA was at first more interested in the power generation potential of fission but, after the attack on Pearl Harbor in December 1941, they grasped the possible military significance and threw vast resources behind a programme to design and build a nuclear weapon. Thus was born the Manhattan Project – perhaps man's biggest and most ambitious scientific and engineering project ever. For over three years 130,000 people, including many of the best scientific minds in the world, worked in secret to produce an atomic bomb. The precious U-235 had to be separated from uranium using massive electromagnetic separators and centrifuges in a great factory set up at Clinton in Tennessee. It was sent to Los Alamos where, in the middle of the New Mexico desert, a vast research establishment was built in total secrecy to design and build the bomb. And with the separated uranium came increasing quantities of a new man-made element, plutonium.

Plutonium came from the exploitation of fission in a different way to the bomb. It had been discovered that if uranium was assembled in a grid with graphite (it came to be called a pile and later a reactor) that the neutrons released in fission were slowed down by collisions with the carbon atoms of the graphite and were then much more likely to cause further fission. The slowing down was called moderation and it meant that a critical mass could be made with un-separated uranium. It could not make a bomb – things happened too slowly – but, because some of the neutrons were emitted after a delay, it could be controlled by inserting rods of cadmium that absorbed some of the neutrons. The results were power (a nuisance at the time) and a whole range of radioactive elements including plutonium (Pu). This element did not occur in nature but the

# GENES, FLIES, BOMBS...

isotope plutonium-239 had the same tendency to fission as U-235. So another bomb by could be made by producing Pu-239 in an atomic pile and then separating it from all the other elements produced. The chemistry of the separation process was not so difficult; the real problem was that the cocktail that came out of the pile was highly radioactive. The intensity of the radiation it gave off was much much greater than had been experience before with radium and x-rays and the material itself was incredibly toxic so it had to be contained to a very high standard. However it provided an alternative route to a weapon to U-235 and with all the uncertainty that existed it was pursued: great plutonium production piles were built beside the Columbia River at Hanford in a remote corner of Washington state. The chemical separation plants were constructed nearby to produce metal ingots of plutonium.

At Los Alamos the challenges were to find ways of creating critical masses of uranium and plutonium quickly (a few kilograms were involved, the size of a grapefruit) and doing this in devices that could be carried and dropped by a bomber. With the uranium weapon the critical mass could be achieved by shooting a lump of U-235 into another lump with a gun. For plutonium this would not work for technical reasons. So a sphere of plutonium-239 had to be compressed uniformly by an encircling explosive charge until it became critical. It was called implosion and it was a massive technical challenge to make it work.

Driven by the ferocious Manhattan Project head General Leslie Groves, by the fear that the Germans would create an atomic bomb before the Allied one was ready (they had discovered fission after all) and, for some, the pure scientific and engineering challenge they faced, both uranium and plutonium bombs were ready in the summer of 1945. The essential simplicity of the gun-type uranium bomb meant that there was high confidence that it would work as planned; the implosion mechanism was more problematic so one of the plutonium bombs was detonated in a test on 16 July 1945 at Alamogordo in the New Mexico desert. Its explosive yield, equivalent to 20,000 tons of TNT (20kt) was almost ten times the size of the Halifax explosion, the largest ever chemical explosion, which had killed 2000 people when an ammunition ship blew up after a collision on 6 December 1917 near the Nova Scotian city. In other words, it worked.[2]

The remaining two bombs were dropped in anger.

### *RADIATION UNITS*

A fairly good indicator of the harm that radiation might cause is how

# THE BOMB

much energy the radiation deposits in a kilogram of the body. This is the radiation dose: if the radiation deposits one Joule of energy in one kilogram then the dose is one Gray – named after a pioneer of radiation studies. Now 1 Gy is really not much energy. It would cause a rise in temperature of little more that 0.001 degrees Celsius. But in terms of effects on our bodies it is very significant.

Since we are going to look first at the immediate effects of the bombs where the radiation dose was delivered over a very short period – a fraction of a second – and spread over the whole body, we will look briefly at the effects that would be found soon after the exposure.

It seems that humans are unable to sense doses of less than about 0.5 Gy but above that level the effects become apparent and increase in severity quite rapidly. The radiation attacks, most significantly, the digestive system and the bone marrow. The first symptoms are generally vomiting and diarrhoea as the radiation destroys the lining of the intestines. If the dose is lower than 2 or 3 Gys then most people will recover but if the dose is more than about 5 Gy then survival becomes increasingly unlikely until at 6 Gy few will live. Death comes from infection and shock after a few days of excruciating abdominal pain and there is little that can be done medically for the patient.

The irradiation of the bone marrow reduces the ability of the body to produce white blood cells and hence to fight infection. Death usually comes from infection so it is possible to save some less-exposed people with appropriate medical treament notably keeping the patient in a sterile environment. Once more, without treatment, around 50% of people will die after a dose of 5 Gy; few at lower doses and more at higher ones. The whole process is rather slower that that asociated with digestive system exposure; people may die several weeks after exposure.

The two syndromes are combined in people who have suffered irradiation of their whole bodies. In cases where just part of the bone marrow is irradiated then the unirradiated marrow will continue to generate white blood cells and these may be sufficient to keep the system fighting infection running, albeit in weakened form. If the radiation dose is spread out over a long period then, even if the entire body is irradiated, rather higher doses may be survived.

### *JAPANESE BOMBS*

Early in the morning of 6 August 1945 the *Enola Gay* dropped the uranium weapon codenamed Little Boy high over Hiroshima and three days later the implosion-type plutonium bomb, Fat Boy, was dropped on Nagasaki from another B-29, *Bockscar*. Large areas of the two cities were flattened by the blasts and the devastating fires that followed destroyed yet more. There is still uncertainty about the numbers who died: estimates range from 150,000 to 250,000. The major killer was the thermal flash but blast must have accounted for a large number of people and the effects of the enormous burst of radiation many more.

The destruction of major hospitals and the death of many medical staff meant that minimal help was available making infection a major problem; at Hiroshima the problems were compounded by a typhoon on 17 September that caused further damage to the infrastructure and slowed recovery.

Within a few days of the Japanese surrender on 2 September, a small team – William Penney, Robert Serber and G T Reynolds – arrived to estimate the yield and survey the physical damage. By 12 October a Joint Commission for the Investigation of the Effects of the Atomic Bomb in Japan had been set up with about 60 US staff and 90 Japanese.

As a result, in November 1946, President Truman issued a Directive requiring a further long-term study. It was to review several specific concerns:

> Cancer, leukaemia, shortened life span, reduced vigour, altered development, sterility, modified genetic patterns, changes in vision, "shifted epidemiology", abnormal pigmentation and epilation.[3]

The Atomic Bomb Casualty Commission (ABCC) was formed the following year with funding through the US Atomic Energy Commission. and it then initiated several studies.

### SURVIVOR STUDIES TO MID-1950s

Miscarriages were more frequent in women who had suffered large radiation doses. By 1952[4] it was also clear that radiation could cause severe damage to the developing embryo (an fact already known from radiotherapy) with the main effect being microcephaly (reduced head size) and associated mental retardation. The early published results showed that there were eleven mentally retarded children born to mothers who had been between 700 and 1200 metres from the hypocentres. Ten of the women had acute effects from the explosions. All of the children had suffered from microcephaly. The most sensitive time for the effect, judging from the Hiroshima data, was between 12 and 18 weeks of gestation.

The development of children who had been exposed to radiation was also studied from the start by the ABCC. Some evidence was found for radiation having affected growth but it was difficult to disentangle the effects of radiation from those due to general conditions and poor

# THE BOMB

nutrition. A follow-up study by Earl Reynolds also found some growth retardation effects but the study was tainted by his plans to photograph naked teenage girls for record purposes. Another approach used x-rays of the children but showed no discernible effects[5].

Early studies aimed at finding any chromosomal damage suffered by survivors were unsuccessful. The procedure adopted – a painful needle biopsy from the testes – gave chromosomes but these were difficult to study. There was even some doubt about the number of human chromosomes. Better techniques were developed and by the mid-1950s damage was being seen.

The first leukaemias were found, not by the ABCC's researchers but by a local doctor, Takuso Yamawaki, who noticed a higher than expected incidence of leukaemia among survivors. All the nine cases he saw in 1949 came from within 1.5 km of the hypocentre. In a study by Folley, Borges and Yamawaki over the next two years[6] 19 more cases were found among nearly 100, 000 survivors at Hiroshima and 10 among a similar number in Nagasaki. By 1953 Moloney and Kastenbaum[7] found 50 cases in Hiroshima with a clear relation to distance from the hypocentre. The increased leukaemia risk was found to be about 1 in 13000 at 2.5 km and beyond; within 1 km it was near 1 in 80. By 1956 , there were 61 cases among the 98,000 survivors followed in Hiroshima.

The possibility of a link between radiation exposure and leukaemias was already strongly suspected from earlier studies of the obituaries of American radiologists[8] and reports of deaths of radiotherapy patients but in 1956 there was a more sinister development when the first evidence of solid tumours was found. The followup to this and its implications were, as we will see, to be the major preoccupation for the rest of the century and beyond.

It was an early priority, given that the genetic effects of radiation had been seen in experiments on plants and animals, to assess the genetic impact. A programme prepared by James V Neel was considered by a special Genetics Conference (Muller attended) and agreed. It was thought that it needed to extend for 10-20 years and not too much should be expected of it – not least because only dominant effects would show up in the first generation. The statement from its meeting in June 1947[9] warned:

> Whether the atomic bombs dropped on Hiroshima and Nagasaki will have detectable genetic effects on the Japanese is a question of widespread interest. The purpose of the present note is to show briefly that (1) many difficulties beset any attempt to obtain a valid answer to this question and (2) even after a long-term study, such

## GENES, FLIES, BOMBS...

as that outlined below, it still may not be possible to determine just how much genetic damage was done at Hiroshima and Nagasaki.

In March 1947 a study was begun, under the direction of William J Schull and and James V Neel, of the impact of the bombs on children born to exposed parents to try to establish whether there were any genetic effects. With the help of midwives and local Japanese doctors, children from over 70,000 pregnancies were examined for malformations between 1948 and 1954. This was done at birth and then, in some cases, 10 months later. In about 12,400 of the cases the parents had been within 2 km of the bombs. No significant effects of radiation could be found.

Data were collected on several other possible effects: neonatal death and infant survival, stillbirths, birth-weight and sex ratio. In no cases were any significant effects of parental irradiation found.[10]

The data collection programme was terminated in early 1954 with the agreement of a second Genetics Conference as the birthrate in Japan fell dramatically. This was largely as a result of the rise of therapeutic abortion following relaxations in the law that controlled it, to allow economic welfare of the mother to be taken into account. The sharp decline in the number of subjects and the unlikelihood of further similar studies producing any meaningful data on genetic effects made it difficult to justify continuing with the massive effort involved.

By the mid-1950s radiation had become a massive public issue as "death dust" from nuclear tests settled from the sky onto every part of the earth and the prospects of a nuclear conflict that could kill millions and leave large parts of the world uninhabitable grew ever sharper.

# 8 FALLOUT

## MULLER 1945-1960

Looking for a job was a painful prospect and a discouraging reality. His left-wing background and his time in the USSR were remembered by potential employers and he had hardly impressed at Amhurst with his teaching skills. But a new element emerged: anti-Semitism. He learned that a major East Coast university had turned him down because he was Jewish. Muller certainly had Jewish ancestors on his mother's side but he and his family had never practised as Jews and he did not feel Jewish. He found anti-Semitism "despicable" but it could not stand in the way of finding a job. He had after all taken a job in Texas where anti-negro feeling was high.

The job he got eventually came through contacts at the Rockefeller Foundation – who had so consistently supported his work. Frank Blair Payne, at the Foundation, had spent some time in Muller's lab in Houston in 1928 and knew his strengths in experimental work. He recommended him to Fernandus Payne at Indiana University. Payne consulted, considered and canvassed and Muller was interviewed and offered a research professorship in the summer of 1945. He privately assured the university President, the inspirational Herman B Wells, that he had never been a member of the Communist Party. He accepted the offer and he and Thea bought a house in Bloomington at 1001 East First Street, a short walk from the campus where he was given space in the basement of ivy-clad Kirkwood Hall.

Here he stayed – except for a year's sabbatical in Hawaii in 1953/4 – until he retired 1964. A move from Kirkwood to nearby Jordan Hall in 1955 gave him more room for the steady stream of research students he nourished and challenged. He continued with his fundamental work on *Drosophila*. Of the 180 papers, books, articles and abstracts published between 1945 and 1961[1] (so averaging a shade under one every two weeks) something like 65 were concerned with the fly. Of the rest there were general works on genetics, but many were on the social issues that had concerned him for most of his adult life. He was much in demand for such things. After all, from 1946 he was a distinguished public figure.

# GENES, FLIES, BOMBS...

## NOBEL PRIZE 1946

Late in October 1946, as Muller prepared to go to Washington where he had been invited to attend a NAS conference stimulated, apparently, by Schrodinger's book, he received a phone call from a New York journalist. The newspaperman told him that he had been awarded the Nobel Prize. Thea danced around the room. Muller tried to calm her down: it could be a hoax, a mistake, a rumour. And so they set off with their secret to Baltimore where they were to stay while Helen was examined for a mysterious illness at Johns Hopkins and Muller travelled to Washington for the conference.

For the first few hours it was difficult in the car as they had a passenger, Dean Riddell, they were to drop off in Cincinnati but after that they were free to discuss the phone call. When they reached Baltimore they checked into a hotel and it was then that they met up with the Raffels. At last there was someone the could share their news with. There was no official confirmation of the award but a Swedish journalist called him at the conference and asked him for a statement. He made none but the word spread around the delegates and many congratulated him. It was only when he returned home that an official telegram was discovered in a pile of accumulated mail.

He collected the prize from King Gustav in a glittering ceremony at the Stockholm Opera House on 10 December – wearing the formal tailcoat he had bought in Edinburgh in 1939. The King, at a skinny 6' 5", dwarfed Muller when he presented him with the Medal and Scroll. In his presentation speech Professor Caspersson of the Royal Caroline Institute summed up Muller's achievements:

> Muller's discovery of the induction of mutations by means of rays has been of tremendous importance for genetics and biology in general.
>
> The foremost instrument of experimental genetics is just the mutations of genes. Thus the whole teaching structure of the Morgan school is based on the utilization of certain spontaneous mutations. Now, when Muller has created a means of simply producing in every laboratory an unlimited number of these otherwise so rare phenomena, it is obvious that genetic research in general must be greatly stimulated thereby. The effect of irradiation is absolutely universal, mutations appear after irradiation within all organisms, from simple viruses and bacteria up to the most highly organized plants and mammals. One of the principal causes of the amazingly rapid development which genetics has undergone during the last two decades is the realization of these technical possibilities.

# FALLOUT

.....

Just this multiplicity of spheres which are affected by Muller's discovery indicates its fundamental character. It is already one of the most important foundation-stones of the complex structure of modern biology, and Mendel, Morgan, and Muller together will always stand out as the creators of the modern science of heredity.

Muller's contribution to its development extends far beyond the discovery for which the prize is now awarded. For more than three decades he has been in the front rank as regards both the scientific work and the eager but inspiring discussions of the results within the field, and these are the most important incitement to future development. He is now more active than ever, and, as the donator wished, the Nobel Prize can now be awarded to a man at the height of his scientific creative power.

In his Nobel Lecture, given two days after the presentation, Muller summarised progress on research that depended on induced mutations. We reproduce here the parts which are most relevant in what follows: the hazards of radiation.

Both earlier and later work by collaborators (Oliver, Hanson, etc.) showed definitely that the frequency of the gene mutations is directly and simply proportional to the dose of irradiation applied, and this despite the wave-length used, whether X- or gamma – or even beta-rays, and despite the timing of the irradiation. These facts have since been established with great exactitude and detail, more especially by Timoféeff and his co-workers. In our more recent work with Raychaudhuri (1939, 1940) these principles have been extended to total doses as low as 400 r [4 Gy], and rates as low as 0.01 r [0.1 mGy] per minute, with gamma rays. They leave, we believe, no escape from the conclusion that there is no threshold dose, and that the individual mutations result from individual "hits", producing genetic effects in their immediate neighborhood.

....

The further the analysis of the genetic effects of irradiation, particularly of the breakage and rearrangement of chromosome parts, has gone, the more does our conviction grow that a large proportion if not the great majority of the somatic effects of irradiation that have been observed by medical men and by students of embryology, regeneration, and general biology, arise secondarily

# GENES, FLIES, BOMBS...

as consequences of genetic effects produced in the somatic cells. The usefulness of this interpretation has been shown in recent studies of Koller, dealing with improved methods of irradiation of mammalian carcinoma. This is too large a subject to digress upon here, but it is to be noted that it has been the analyses based in the first place on genetic and cytogenetic studies of the reproductive cells, as shown by subsequent generations, which are thus helping to clear the way for an understanding of the mechanism by which radiation acts in inhibiting growth, in causing sterilization, in producing necrosis and burns, in causing recession of malignant tissue, and perhaps also, on occasion at least, in inducing the initiation of such tissue.

.....

In this situation we can, however, draw the practical lesson, from the fact of the great majority of mutations being undesirable, that their further random production in ourselves should so far as possible be rigorously avoided. As we can infer with certainty from experiments on lower organisms that all high-energy radiation must produce such mutations in man, it becomes an obligation for radiologists - though one far too little observed as yet in most countries - to insist that the simple precautions are taken which are necessary for shielding the gonads, whenever people are exposed to such radiation, either in industry or in medical practice. And, with the coming increasing use of atomic energy, even for peace-time purposes, the problem will become very important of insuring that the human germ plasm - the all-important material of which we are the temporary custodians - is effectively protected from this additional and potent source of permanent contamination.

The award gave his views additional weight in the debates about atomic weapons.

## *CROSSROADS*

Three tests of implosion-type plutonium bombs were planned for 1946 with the stated intention of seeing the effects on naval vessels and personnel (represented by pigs and shaven goats, by guinea pigs and rats). The first test, Able[2], took place on 30 June (after a six week delay ordered by the President perhaps as a result of a Russian accusation that the US was "brandishing" the bomb at a delicate time in international relations) and was something of an anticlimax. *The Times* reported the following day[3] that a bomb had been dropped over a target fleet assembled in Bikini lagoon but the "Target Ship Remains Afloat", "Cloud Gone in 90 Minutes", "No Ships Sunk", "Aircraft Unaffected". There was no tidal wave and the

# FALLOUT

live coverage of the event broadcast by the BBC Light programme even failed to register the explosion. The target ship, the battleship *Nevada*, had been painted orange and was relatively undamaged. This proved to be because the bomb had exploded about two miles off-target.

Illustration 7: Crossroads Baker

Closer inspection showed that the damage had been more than it appeared from a distance. It seemed unlikely that the crews of many of the vessels could have survived – although goats were seen still munching hay on the decks of some of the boats. The *Times* headline of 2 July was "Holocaust at Bikini." But many of the press seemed unimpressed and the *Manchester Guardian* ended its 2 July report: "The reaction among correspondents covering the test was that it was 'disappointing'.". Many of the press left before the next explosion, designated Baker, on 25 July.

Baker was a similar device to Able, with a yield of about 20 kiloton, but this time it was detonated 30m below the lagoon surface. The result was dramatic even by the standards of these awesome weapons. A vast spout of water shot up thousands of metres surrounded by the domed

# GENES, FLIES, BOMBS...

condensation cloud which quickly expanded and evaporated. Then the water column was seen below the boiling and expanding mushroom cloud. The 26,000 ton battleship *Arkansas*, its hull vertical, seemed to climb the column. And then a great turmoil of highly-radioactive spray and 20 metre waves swept outwards and engulfed the ships and million tons of radioactive water were thrown into the sky to fall back into the lagoon. This time the explosion was heard in Britain on the Light programme.

Many of the ships in the target array had been severely damaged and a few sank soon after the test but the visible evidence of the explosion soon faded away, the lagoon settled and the cloud drifted off. King Juda of Bikini, brought to witness the test, could report back to his exiled subjects on Rongerik that the island was still there and the palm trees were still standing. But Baker had introduced a new element into the debate: radioactive contamination. The great cloud of spray that had swept across the ships had left many of them so contaminated that they could not be cleaned; the magnificent German battlecruiser *Prinz Eugen*, that had steamed into the Atlantic in 1941 with the battleship *Bismarck*, sank several months after the test, too radioactive to repair.

The Japanese explosions and Able had been air bursts so had not generated vast amounts of contamination. In Japan the devastation had been so obvious and complete, so many had died from blast and heat and the prompt irradiation from the bomb, that the contamination had hardly been noticed. The Able test had contaminated a few ships but nothing more. This was different. The clean-up would be so difficult that the third Crossroads test, Charlie, was cancelled.

On the anniversary of Crossroads the American magazine *Life* published a special edition on the tests. Its generally upbeat tone was spoilt by the revelations of Stafford Warren[4], the former chief of medical services for the Manhattan Project and the head of radiological safety for Crossroads, who pointed to the clinging radioactivity of the target ships and to the persistent radioactivity in the algae and fish of the lagoon. He concluded that a similar bomb, exploded in New York harbour might kill two million people.

Another Bikini voice, in 1948, spoke of the dangers of radioactivity. David Bradley had served in Warren's radiological safety unit as a physician and wrote *No Place to Hide* about his experiences during the Crossroads tests. As the title suggests, the main theme of the book, leavened with stories from his journal, was that use of the bomb would lead to a future with man crippled by the presence of ineradicable and inescapable radioactivity. The book was in the *New York Times* best-seller list for ten weeks and sold a quarter of a million copies in little more

# FALLOUT

than a year. Radioactivity was now ingrained in the public mind – as it had been in the hulks of Bikini lagoon.

There were no tests in 1947 but three in 1948. The only test of 1949 was "Joe 1" on 29 August 1949 and it stunned the world: it was undertaken in Semipalatinsk, in the remote Kazakhstan region of the USSR. It showed that the Soviets were much more advanced in nuclear weaponry than the West believed. Britain's MI5 already suspected that there was a spy feeding atomic secrets to them and not long after the test they identified him as Klaus Fuchs, the head of the Theoretical Physics Division at Harwell[5]. Fuchs, a pre-war refugee from Germany, had worked on several aspects of the fission bomb, notably the implosion technology, as part of the Manhattan Project. He also contributed to the fusion weapon development – and fed information steadily on all of it to the Russians from 1941. He returned to the Atomic Energy Research Establishment at Harwell to Britain in 1946 and was confronted with evidence of his spying in December 1949. He denied everything initially but confessed in January 1950. He was convicted and sentenced to 14 years in gaol, released after serving nine years when he returned to East Germany.

There were no tests in 1950 but then 18 in 1951. The USSR conducted two at Semipalatinsk and the USA tested four devices in the Pacific and 12 at their newly established Nevada Test Site.

From now on, until testing ended completely, the Americans used the Nevada test site for smaller devices; smaller is of course a relative term and some of the detonations had yields similar to that of the Japanese bombs. The early tests were almost all above ground; some devices were exploded on towers, some dropped from aeroplanes. Although mainly for weapons development, they were used to study radiological effects on animals. Troops were exposed to some explosions in exercises. A few other tests took place around the continental United States.

The tests at the Nevada test site were perhaps a source of pride as much as dread. *Life* magazine had a front cover that showed the buildings of Las Vegas silhouetted by a dawn flash. Tourists came to the atomic city for the regular displays. The Ranger Fox explosion of a 22 kt weapon 400 metres above the ground on 6 February 1951 at Frenchman Flat broke a few shop windows in Las Vegas (they had been warned) and rattled more in Los Angeles, 300 miles away. The explosion was seen in Boise, Idaho, 400 miles north of the test site. Translated into Europe this would have meant an explosion in central England shaking Paris and being seen in Copenhagen. There was no significant fallout beyond the test site.

Later tests were more controversial. The Upshot-Knothole Harry test of

# GENES, FLIES, BOMBS...

May 1953 sent a cloud of fallout over St George, Utah where, according to locals, it lingered causing lesions on animals. Later there were claims of high radio-iodine levels and increased levels of cancer in the area resulting from this and other tests. The "Downwinders", as they called themselves, pursued the AEC and US Government for many years for an admission of guilt but the court cases fizzled out. There was finally recognition of the problem, an admission of guilt and, for some, compensation when the Radiation Exposure Compensation Act came into force in 1990.

Just before the "Dirty Harry" test the Simon test resulted in detectable fallout in far-away New York State. Rain seems to have washed some of the cloud onto the streets of Troy, 2300 miles away, where it was detected on 26 April by Geiger counters in the Rensselaer Polytechnic chemistry laboratory of Herbert Clark. Clark, who had worked on the Manhattan Project, investigated the radioactivity with his students and wrote an account for Science, concluding that it was indeed fallout but at a very low, non-hazardous level.

With the return of the testing to the continental USA the AEC set up a network to record fallout across the country.[6] One hundred and twenty one Weather Bureau stations around the continental USA (excluding Alaska) collected settled dust on gummed paper and sent it daily to New York for analysis. At some stations air was drawn through filter papers and these were sent too.

In addition to this network the AEC also deployed mobile monitoring teams between 200 and 500 miles of the test grounds with similar equipment whenever a test was planned. The data given were quite full but not particularly easy for the layman to interpret. It was clear enough that there was easily detectable fallout from the Nevada tests – and it was more marked than the fallout in the US from the Pacific tests. Higher readings were found in a wide belt stretching east across the northern states as far as the east coast. The authors tried to put the internal hazards in perspective by comparing strontium levels with naturally-occurring concentrations of radium in soil: the fallout hazards were much lower than the background ones. The gamma radiation levels from the fallout were no more than would be received in 25 days from natural sources.

The UK started testing on the Monte Bello Islands off the north-west coast of Australia in October 1952 before moving most of the testing to Emu Fields and Maralinga in the south-west of the country. The yields of weapons tested were generally in the region of 20 kt.

So by 1953 the three nuclear weapons nations of the time had carried out a few dozen tests of atomic weapons and the fallout issue was on the

# FALLOUT

table, largely as a result of the contamination reported from the USA's tests. The fallout from weapons testing was no longer dropping, apparently harmlessly, into the distant and empty Pacific but was raining on Americans in their homeland. Quite how much real concern it caused nationally it is difficult to say; ironically perhaps it was an event in the Pacific that was to change perceptions dramatically.

## CASTLE BRAVO

There were limits to the yield of an atomic weapon and the possibility of a much more powerful device had been discussed during the early days of the Manhattan Project. The "Super" (or hydrogen bomb) would use a fission weapon to trigger a fusion device[7] with much more energy being released. Not everyone agreed that such a weapon was a practical proposition and hardly anyone thought it could make a difference to the war going on at the time. Any serious work was postponed until after that had been won.

Edward Teller was by far the most persistent advocate of the hydrogen bomb throughout and after the war. However, the doubters – doubting both its practicality and the need for it – dominated the scene until "Joe1". Suddenly and shockingly the USSR was an atomic power. Determined to regain the lead, President Harry Truman quickly announced that work would start on the H-bomb. Although doubts even now persisted about feasibility, within little more than a year Teller and Stanislaw Ulam had a design that seemed likely to work and have a yield in the megaton range – so more than 30 times that of the more powerful A-bombs of the time. The design was confirmed in the Ivy Mike test of 1 November 1952 with a yield of 10 Mt.

The Castle Bravo test of 1 March 1954 was the first one of a practical hydrogen bomb. It took place on a platform just west of the small island of Namu which formed part of Bikini Atoll. At 15 megatons it was more than twice as powerful as expected. This was not too significant but it did mean that some of the instrumentation was destroyed – and it was something of a surprise to the firing team located just 20 miles away. Much more serious was the fact that the winds behaved not quite as predicted and swept the fallout not out towards the empty ocean but towards the islands of the Rongelap and Rongerik atolls, over 100 miles away, forcing the evacuation of the 236 Marshallese inhabitants and of 28 US weather personnel on the islands. The fallout also rained on the Japanese vessel *Daigo Fukuryū Maru* (Lucky Dragon Number 5), fishing for tuna 85 miles away. The crew saw the glow of the explosion and heard the rumble that followed a few minutes later. Then a few hours later the fallout arrived, like snow, covering the 23 crew and the boat. They cleaned it off, as best

# GENES, FLIES, BOMBS...

they could, with their hands. Most of the crew suffered some effects of radiation and were hospitalised on their return to Japan on 14 March. Their catch of tuna was destroyed along with much suspect fish from other boats and many people stopped eating fish. The price of tuna fell by 50%. The entire Japanese nation was enraged: they understood the horror of nuclear weapons, there were Japanese radiation casualties again and it seemed that their fishing industry might collapse.

The USA response was poorly judged. At a press conference on 31 March President Eisenhower admitted that "something unexpected must have happened" and introduced Lewis Strauss, the Chairman of the AEC, who suggested that the injuries to the Japanese fisherman were likely due to the chemical nature of the fallout[8]. Two US experts sent to Japan to assess the injuries of the two most-exposed fishermen said that "they would be all right within a month". A Senator present at the test said the injuries were "considerably exaggerated." The boat, it was said, must have been fishing in the declared danger zone. As for the weather personnel who had been exposed he said none had burns; "The 236 natives also appeared to me to be well and happy."

There were reassurances from Strauss about the fish. No contamination could reach Japanese home waters he said and, with the possible exception of the Lucky Dragon catch, all the fish being destroyed were actually safe to eat. Indeed, in a more than usually desperate attempt at evasion, he pointed out that, at some times of the year, the fish caught around the Marshall Islands were poisonous anyway.

One of the fishermen died a miserable death that September from complications that might, it has been claimed, have arisen from the treatment intended to save him. The 2 km wide bomb crater, readily visible in satellite images of a Pacific jewel, will be a enduring reminder of the USA's largest nuclear explosion.

## CONSEQUENCES FALLOUT

In the months following the disaster of Castle Bravo the AEC worked up a short rather non-technical document that aimed to explain some of the issues about hydrogen bombs, particularly fallout, to members of the public.[9]

It explained that 0.25 Gy would cause temporary effects on the blood, 1 Gy over a short period would cause radiation sickness and 4.5 Gy over a day or so might be fatal to about half the people exposed. If the exposure were delivered over longer periods, the effects would be less.[10]

The Castle Bravo test had produced fallout ("somewhat adhesive" slaked

# FALLOUT

lime produced from the coral) that spread in a cigar-shaped pattern 140 miles long and up to 20 miles wide. Anyone within that area who did not take precautions would have been very likely to die. The radiation dose on the north-western part of Rongelap (100 miles from the detonation point) was revealed to have been 23 Gy in the 36 hours after fallout began. Even at 160 miles distant, on the axis of the cigar, one half of the population might die from fallout. The report explained how simple actions could increase chances of survival quite dramatically. So this part of the report was directed towards survival in wartime.

But concerns about fallout from tests were addressed too. People, it was acknowledged, were worried about radio-strontium and radio-iodine[11] but they need not be. The AEC been studying the levels of these in the environment for three years using methods that were so sensitive that adequate warning could be given of any dangers well before they materialised. The levels were, currently, insignificant outside the test areas themselves.

There was uncertainty about heritable genetic damage:

> At our present stage of genetic knowledge there is a rather wide range of admissible opinion on this subject.

However, the total exposure of a typical US resident to date from all the explosions that had taken place amounted to about 1 mGy. This was, the report continued, about one-hundredth of the exposure received from natural background over a person's reproductive lifetime and about the same as that from one chest x-ray. At such a low level it would not "seriously affect the genetic constitution of human beings."

### *CONSEQUENCES WAR*

After reading a statement to the President's press conference, Strauss invited questions. He was asked if there was any limit to the power of a thermonuclear weapon. He replied that it could be made "as large as you wish" and added "It could destroy a city."

A reporter asked if the bomb could wipe out New York.

> Without a moment's pause Admiral Strauss answered that the hydrogen bomb could "destroy the metropolitan area of New York." Stunned and unbelieving the reporters asked him to repeat his answer. Again the grim statement came, without any change in it's stark emphasis.[12]

The Guardian's headline was "UNLIMITED POWER OF H-BOMBS, One Could Destroy Whole of New York."

## GENES, FLIES, BOMBS...

The White House later issued some clarification. Strauss meant that the bomb could put a city as large as New York out of commission but not level it to the ground. Whatever, the fallout from Strauss's remarks proved, like that in the Pacific, somewhat adhesive.

So the Castle Bravo disaster had two linked but different effects: it made people aware that fallout was not just some harmless scientific novelty that made Geiger counters click but was potentially, even in tests, a widespread threat to health and it moved the focus from the blast and heat of a bomb in wartime to the large swathes of the land that would be covered in the deadly dust.

# 9 RADIATION HARMS

## *RADIATION STANDARDS*

The history of radiation protection is a convoluted technical one[1] and here it is restricted to the basics.

Individual nations had set up their own bodies (or they had been set up) to make recommendations on protection of x-ray and radium workers, with of course a medical bias. Starting in the 1920s, and led initially by the British, international standards began to emerge for work with x-rays and, later, radium. These set design standards for for x-ray enclosures and operations (including on working hours and holidays) but no exposure standards as such were set. It was not until 1934 that the first radiation exposure limits were recommended by the International X-ray Protection Committee (IXRPC):

> The evidence at present available appears to suggest that under satisfactory working conditions a person in normal health can tolerate exposure to X rays to an extent of about 0.2 international röntgens (r) per day.[2]

In keeping with the aim of not becoming involved in too much detail about radiation quantities and units we will translate all units of the time into into absorbed dose in Grays and related quantities. Thus 0.2 international roentgen becomes 2 mGy.

The problems in setting such standards were at least two-fold. First there were those associated with measurement (which this account will tend to ignore) and second deciding what level the acceptable dose should be set at.

The early effects had been horrendous for some early radiologists, a number of whom had a series of amputations of fingers and hands before succumbing to disease. Some developed cancer. The cancers that did arise were not recognised as a sinister stochastic threat (that might arise after *any* radiation dose however small) but rather as a consequence of the terrible injuries and it was thought that all these effects could be avoided if only doses could be kept below levels that caused visible effects. The most modest effect seemed to be erythema – the reddening of the skin after exposure – so this became a benchmark. Keep an adequate amount below this and the body would tolerate the exposure and there would be no adverse effects. This threshold approach persisted until after the Second World War and the main concern was establishing what the "tolerance dose" actually was.

In fact the 2 mGy per day had a rather short life internationally: in 1938 the influential US Advisory Committee recommended reducing it by 50% to 1 mGy per day[3] and this was subsequently taken as the standard for workers in the Manhattan Project.

Genetic harm played no part in the considerations but it had apparently been discussed by the US Advisory Committee in 1940. However, with the preoccupations of war it was not taken up further at the time; indeed the file that recorded the discussions seems to have gone missing.

> It [the US Advisory Committee] had in 1940, actually, pushed the idea of genetic effects, backed away from those somewhat in the late 1940's, but it began to appreciate that it wasn't going to go away, so Muller was brought in.[4]

The US Advisory Committee was transformed after the war to reflect the enormously increased scope of radiation work generated by the Manhattan Project and renamed the National Council on Radiation Protection and Measurement (NCRP). A structure with a main committee, an executive committee and no less than 15 sub-committees was set up. Taylor, the Chairman of the Advisory Committee was voted Chairman (for an indefinite period) of the new body and Hermann Muller joined Subcommittee 1: *Permissible doses from External Sources*.[5]

The International Commission on Radiological Protection (ICRP)[6] originated, under a different name, in 1928 but it was really only after the war, in 1950, that it began to play a major role. It continues to this day as the international focus for much work on the safety of radiation and is, particularly, the recognised body that recommends standards to national authorities.

An insight into the preoccupations of standards setters in the immediate postwar period comes from the minutes and papers of the Tripartite Talks involving the USA, the UK and Canada in the late 1940's and early 50's[7]. These were mainly about the problem posed by the generation of so much new radioactive material and the possibility that it might be ingested; the problem of internal dose assessment. However they did debate two issues of relevance here: the dose limits and the risks of genetic effects.

# RADIATION HARMS

For the dose limits there was much discussion about the apparent differences between US and UK proposals. In 1946 the NCRP had proposed a limit of 3 mGy/week (so more or less halving the wartime value) and this was agreed at the Talks in 1949 with the understanding that it might be represented in slightly different ways in different countries. It meant careful wording of documents but this was agreed and the ICRP recommendations of 1950 reflected that. The essential reason for the tightening of standards was stated by Gioacchino Failla in his Preparatory Report for the Talks: large scale experiments on rats and mice had suggested that the margin of safety on the tolerance dose of 1 mGy/day was not as large as thought. He also said that a revision downwards was required on genetic grounds and suggested a limit of 3 Gy in a lifetime. It was decided however not to go ahead with such a limit for the time being; enough protection came, in practice, from the other limits proposed.

The question of exposures came up in a very specific context that illustrated the cracks growing in the confidence surrounding the tolerance approach. The discharge of wastes containing traces of plutonium took place from AERE Harwell into the Thames, contaminating this important source of water for the capital and there was some discussion of the permissible level of radioactivity in drinking water. It was decided that this should be set to ensure that exposures were no more than 1/100th the tolerance dose for workers for the reason that it was not possible to be sure that all effects of radiation were actually threshold ones: some might be linear. While this concern may have been prompted by the knowledge that genetic effects were not threshold the example actually given was bone tumours. The practical conclusion was actually that the UK might push this reduction publicly but that the USA were not willing to commit to addressing a problem they did not currently have (or possibly were not prepared to admit).

It should be said that, while there may have been some doubts about the tolerance approach and the threshold notion, they were ingrained in the approach the Talks took to the problem they spent most time on: the effects of internal radioactivity. Here the key concept was the critical organ (or tissue) and the protection aim was to limit the dose to that organ to below a threshold.

There was rather little consideration of genetic harm over the series of meetings. However, at one meeting there was discussion on doubling dose, the dose required to double the spontaneous mutation rate. The UK team said that the work at Edinburgh, while not too far advanced, already showed that the doubling dose could hardly be less than 0.7 Gy. The US team mentioned the work of D R Charles at Rochester saying that the data were still being analysed (full publication was, in fact, a decade away) and the 0.5 Gy figure for the doubling dose they had quoted was still tentative. They offered to send information about the new work being started at Oak Ridge by William Russell to the Edinburgh group. In the discussion on the doubling dose it became clear that there was considerable uncertainty about the spontaneous rates of mutation to be used with the doubling dose. The problems of translating any results from *Drosophila* and mice to humans were also mentioned; both induced and spontaneous rates differed between species as well as between loci in any one species.

The conclusions of the Tripartite talks were generally reflected in the ICRP 1950 recommendations[8]. The general effects on the body and particularly those on the blood-forming organs (such as anaemia and leukaemias) were seen as the most dangerous. The risks of malignant tumours was also considered as were those of genetic effects. The maximum permissible whole-body exposure to external x-radiation was reduced from the 10 mGy/week of IXRPC in 1934, now considered "very close to the probable threshold for adverse effects", to about 3 mGy/week.[9]

## EQUIVALENT DOSE

At some point the reader needs to be aware of a complication to the whole idea of radiation and its effect on people. Different sorts of radiation (x-rays, gamma rays, beta particles, alpha particles and neutrons) interact with tissue in different ways.

X-rays and gamma-rays are both electromagnetic radiation but they differ in their wavelengths – and so in the energy their photons carry. Their interactions with matter are rather different but they both result in high-energy electrons that skip through tissue, ripping electrons from atoms in their track (leaving them ionised) because of their electric charge. Beta particles are just high-speed electrons so they have a very similar effect.

Alpha-particles are much heavier beasts than electrons and they plough through matter leaving a short thick track of ionised atoms behind them. The damage they cause is much the same as that caused by electrons except that the ionised atoms are much close together. This closed packed ionisation seems to be much more damaging to living cells than the spread out kind – that much had come from experiments starting before WW2. As a result it's not just the absorbed dose that matters in estimating the risks but the nature of the radiation. As a general rule alpha-particles are about 20 times more dangerous per Gray than electrons – and therefore than x-rays and gamma-rays.

# RADIATION HARMS

The interactions of neutrons with matter are extremely complicated and vary with the energy of the neutron. However, many of the products are densely ionising particles and, like alpha-particles, cause more damage than the absorbed dose in Grays would suggest. The multiplying factor can, depending on their energy, be between one and 20: an indicative average value of ten seems about right.

So multiplying the absorbed dose in Grays by this weighting factor (1 for x-rays, 20 for alpha-particles and say 10 for neutrons) gives a better indication of the risk from the different radiation. It is called the equivalent dose[10] and is measured in Sieverts (Sv).

From now on we will generally express radiation dose in Sieverts.

Equivalent dose is sometimes referred to by an older name: "dose equivalent".

## GENETIC LOAD

A good point to start is the prevailing view of the role of mutations in disease expressed in the J AM Med Assoc in 1947 — and quoted at the start of *Our load of mutations*[11]:

> While a few students of heredity have maintained that hereditary diseases and certain congenital anomalies and malformations in man not infrequently may arise from mutation, although unable to present any indisputable evidence in proof of their hypothesis, until recently the prevailing view has been that mutation as a direct cause of disease is extremely rare and of little practical significance.

Muller had been convinced for some time that the long-term genetic risks associated with radiation had been unrecognised and had spoken against injudicious use of x-rays. He now began to find a potent vocabulary to argue the point.

Muller introduced the idea in his Pilgrim Trust Lecture of November 1945[12] delivered at the Royal Society in London after an Atlantic crossing on the *Queen Mary* through seas still haunted by feral mines. His words were already those of polemic: "one must recognise...the inexorable rule that practically every 'small' and non-lethal mutation requires finally a genetic death" and "for each mutation, then, a genetic death".

In *Mutational prophylaxis:*[13]

> ... a detrimental gene giving only a one per cent reduction in the chances of survival would tend to go on down until it had been able to "show" itself in a hundred individuals, on the average, before it finally killed one of them or prevented it from reproducing, and so

> died out.
>
> This conclusion is very important, because it shows that ultimately every small harmful mutation is just as bad, if not worse, in its final effects on the population, than a fully lethal mutation.

In this paper he noted the earlier work of J B S Haldane in 1937 that the decline in average relative fitness of a population resulting from mutations is proportional to the rate of mutation and *not* to the degree of harm caused by each one. Fitness in this context means a combination of viability and fertility and is equal to the number offspring expected to be left by an individual and to attain reproductive age. Relative fitness is the fitness relative to some standard genotype. The somewhat surprising result applies to dominant, recessive and partially dominant genes. It is essentially an expression of the principle that a mutant gene will continue to exist until its carrier fails to reproduce. If the mutation is lethal then it will die out in the first generation; if it is merely detrimental it will persist for many generations and affect many people before eventually there is a death or a failure to reproduce.

In *Our load of mutations* in 1950 he acknowledged the work reported by CH Danforth at the second eugenics conference in 1921

> From Danforth's observation that all detrimental mutant genes, of whatever grade, tend to equilibrium frequencies at which their extinction rate simply equals their rate of origination by mutation, Au, it follows very directly that the grade of detriment occasioned by a gene when it manifests itself in an individual has no influence upon the amount of genetic death it causes in the equilibrium population. For its death rate depends only on its mutation rate. And since the death rate is a kind of index of the total damage which the gene occasions, it also follows that, paradoxically, the grade of detriment caused by a gene in the average individual in which it manifests itself is not correlated with the total amount of damage it does in the entire population. All this results from the fact that a less detrimental impairment accumulates to a compensatingly higher equilibrium frequency than a more detrimental one of the same mutation rate.

This dramatic expression was, as we will see, open to technical criticism but it certainly made it clear that in Muller's view mutations introduced a burden that would persist for a very long time and about which nothing could be done.

# RADIATION HARMS

## *THE LIMITATIONS OF MULLER'S IDEA*

The genetic death idea relates to what Muller calls harmful or detrimental. One might assume that he means that these would include traits that may be distressing or incapacitating but have no effect on reproduction. However, by harmful or detrimental he has to mean that the mutation has some effect, however small this may be, on reproductive capacity or, more generally, on fitness.

Muller's use of the genetic load concept in relation to radiation is summarised by Wallace as:

> In Western civilizations, populations maintain themselves (and even expand) because each couple, on the average, leaves slightly more than two children. This reproductive pattern leads to as many or more children than there are parents. If the mutational load in man is 0.20, the average number of surviving children per pair must represent only 80% of those that might have been born; at least 2.5 children must be conceived per couple on the average if (1) only 80% survive and (2) the population does not dwindle with each passing generation. If levels of man-made radiation were to increase until artificially induced mutations had effectively doubled the present-day "spontaneous" mutation rate, man's genetic load would increase eventually to 0.40. In this case, only 60% of all children conceived would actually survive. We are producing an average of 2.5 children per couple in order to maintain the population as things are now; with the increased genetic load we would produce an average of only 1.5 surviving children per couple. This low rate of reproduction would lead to the extinction of the human race.[14]

As Wallace points out, human population is not limited by capacity to reproduce but rather by choice and the populations of some countries with very high infant mortality actually increase in size.

When he came to write the summary of the BEAR committee report of 1956 (of which, more later) Warren Weaver would write:

> One way of thinking about this problem of genetic damage is to assume that all kinds of mutations on the average produce equivalent damage, whether as a drastic effect on one individual who leaves no descendants because of this damage, or a wider effect on many. Under this view, the total damage is measured by the number of mutations induced by a given increase in radiation, this number to be multiplied in one's mind by the average damage from a typical mutation.

# GENES, FLIES, BOMBS...

> Measuring total damage in terms of the number of mutations does indeed necessarily involve this concept of the average damage from a typical mutation, and some geneticists find this concept difficult and illusive. They would point out that mutations may be grouped in classes that differ, on a subjective scale, many thousand-fold in the amount of damage per mutation. As examples they would cite a mutation which results in very early death of an embryo (which might cause very little social or personal distress), and a mutation which results in severe malformation to a surviving child, (which would cause very great personal distress and which clearly involves a social burden).[15]

It seems that Muller was trying to bundle together two effects of mutations: the misery they might cause to individuals who inherited them and the risk of death or infertility they might bring in either hetero- or homozygous forms. A mutation which leads to relatively minor incapacity but has little effect on fitness might would cause distress for many generations and many individuals. One that caused major incapacity and a serious decrease in fitness would be distressing for far fewer people as it would be eliminated much earlier.

Wallace would write in 1991:

> Unfortunately, the account [in one of his earlier works] proceeds from my acceptance that a load (genetic or phenotypic) harms a population, to a later belief that it has little or no bearing on a population's well-being, to my present feeling that a phenotypic load (which may or may not have a genetic basis) provides the means for the culling of excess individuals, thus avoiding overcrowding and increasing the probability that a population will persist through time.[16]

The genetic death concept had other limitations:

> The trouble with this vision [as expressed in Warren Weaver's summary] is that it is difficult to translate into societal costs and actual human suffering. For this reason the approach has fallen into disuse.[17]

## *THE MOUSE*

With the rediscovery of Mendel and the enormous interest in evolution the natural question was whether Mendelism applied to man. Mendel himself had wanted to experiment with mice but was forbidden by his abbot who was revolted by the idea that one of his flock should share accommodation with an animal, caged or not, for purposes of procreation.

# RADIATION HARMS

So Mendel had to content himself with peas. Scientists did experiment with other higher animals: chickens, guinea pigs, pigeons and rats to name a few. But the mouse established itself quite quickly as an important experimental animal – and in some degree a model for man.

One advantage of the mouse was that it had been bred, mainly for different coat characteristics, for a long time (the Japanese had collected mutants for coat colour for several centuries), and so there was a variety of characteristics already available. By 1902 Cuénot had already shown that coat colours were inherited in a way consistent with Mendel's ideas and by 1905 he recorded the yellow coat lethal mutation (which was explained by W E Castle and Clarence Little of the Bussey Institute at Harvard 1910). The year before he found the first multiple allele – not to be found in *Drosophila* for another nine years.

It was Leo Loeb who, at the very beginning of the century, propelled the mouse on the road to laboratory dominance when he found that tumours could be transplanted and would grow in Japanese waltzing mice but common mice were quite resistant. For some years it was thought that the inheritance of this resistance was non-Mendelian. Cancer was then, as now, a major preoccupation of medical researchers and these results were considered important insights into the disease. Clarence Little built up a stock of inbred mice specifically to test this and, with Tyzzer, demonstrated in 1916 that it was, in fact, Mendelian.[18]

The cancer studies with the little animal were to continue for decades and came to include work on spontaneous neoplasms and led to the development of numerous special strains of laboratory mouse – with Clarence Little taking a leading role. It provided a model animal, not too removed from man, which was easy to keep, keen to breed and agreeably fertile. Paigen summed up the role of the mouse in cancer studies:

> Cancer was the driving force that carried mouse genetics through its first 5 decades and greatly influenced the development of the mouse as a genetic system. The pressure to solve an important medical problem resulted in the creation of a new experimental system that was to have far wider application in the years to come. For the cancer problem itself, the eventual outcome proved to be one of those recurrent ironies of scientific history. While the study of spontaneous neoplasms led to the discovery of retroviruses and oncogenes and has brought us to the brink of a deep understanding of the biological basis of cancer at a molecular level, the studies of tumor transplantation, which started it all, had no significant impact on our understanding of cancer. Rather, in leading to the increasing numbers of genetic variants with altered enzyme activity,

until a fairly comprehensive picture of the studies inadvertently initiated the description of a molecular complex central to the operation of cellular immunity.[19]

There were two people who made significant impacts on with studies of the genetic effects of radiation on mice. George Snell, who had worked with Muller as a post doc andobtained a post at the Jackson Lab in 1935. His work for the first few years was on radiation effects and he found a few mutations in offspring of irradiated mice when they were autopsied. The numbers of mice examined were fairly small but it was a significant finding. Snell moved on at the end of the thirties and devoted most of his time after that to another area of research that had been a preoccupation at Bar Harbor: the genetics of tissue compatibility. The work he did led to him being awarded a Nobel Prize in 1980 for the work – referred to in the last sentence of the quotation above.

A contribution to radiation genetics came from Paula Hertwig, daughter of Oscar Hertwig and niece of Richard Hertwig (who had developed the three tissue layer theory of embryo development) working in Berlin. The first woman in Germany to obtain a post-doctoral qualification, she struggled to find a permanent research post and teaching work She found several visible recessive mutations after irradiation of male mice in work published between 1939 and the 1950s. The sample was not very large but her results in 1939 and 1941 were probably the first that could be used in a quantitative assessment of the genetic effects of radiation.

So when concerns about radiation effects on man the mouse was the experimental animal of choice – in its millions.

A large study of the genetic effects of chronic irradiation of mice was carried out at the University of Rochester between 1943 and 1950 by D R Charles, first under the Manhattan Project and later the Atomic Energy Commission. There was a preliminary report in 1950 and a (posthumous as far as Charles himself was concerned) summary report in 1961.[20]

According to the summary paper, 9000 mandays had been spent on the project and a total of 400,000 mice had been bred. The core study involved the chronic irradiation of male mice at rates between 1mGy and 100 mGy per day (for how long is not clear) and examination of their offspring. In total 136 males were irradiated and there were 51 controls. All 12,000 offspring were raised to maturity, females were bred to see if there were any effects of radiation on their fertility and then all the offspring were autopsied for abnormalities. It was acknowledged that the study would have detected only dominant mutations and the 1961 paper summarises the results as:

# RADIATION HARMS

Genetic damage to the immediate off-spring of the irradiated mice was detected in the forms of increased juvenile mortality, increased occurrence of rare morphological abnormalities, and progeny tested mutations affecting both morphology and fecundity. Several factors, notably sample sizes, render accurate quantitative interpretation of the experimental results difficult, but some damage appeared demonstrable at even the lowest levels of treatment.

A perhaps disappointing result after so much work. One of Charles's team, Arnold Grobman, published *Our Atomic Heritage* in 1951. It gives some description of the experiments and attempts to draw conclusions about their implications for humans.[21]

There had been even earlier experiments with the mouse and possible radiation-induced mutations: Little and Bagg in 1923 might have anticipated Muller's discovery in finding mutations in the offspring of irradiated mice but they had too few mice and too small doses. They found just two abnormalities – and there was one in the control group. Little's largest technical contribution to mouse was probably the establishment just after World War of two strains of lab mouse that are still important today: the dilute brown (DBA) and black (C57BL). The sub-strain 6 of the black accounts for more that half the mice used in experiments around the world. It has many technical advantages with a personality that shares a tendency to bite with an enthusiasm for alcohol. However Little's main legacy is possibly organisational. After early success as an academic administrator, becoming the youngest ever president of the University of Maine at 33, and a spell as President of the University of Michigan – which was described by the University's historian as "stormy" and his approach characterised as "brilliant but tactless" – he resigned in 1929 to set up the Jackson Laboratory at Bar Harbor, Maine. The year 1929, given the Wall Street Crash and the Depression that followed, was a bad time to start a research institute that depended upon private funding and the Lab struggled to survive on donations, what government money Little could muster and the sales of its stock mice. The number of researchers had grown from seven to 18 when, in October 1947, the Laboratory was burnt down in a fire that consumed most of the research records and many of the animals.

This was to prove significant for our story: one of the researchers was an Englishman, William Russell, recently divorced from a co-worker and even more recently married to another one. Bill and his new wife Liane (known generally as Lee) had been encouraged to move away from Bar Harbor by the scandal of their hurried nuptials and had been offered jobs together at Oak Ridge. At the time of the fire they were waiting for their

# GENES, FLIES, BOMBS...

FBI clearances; a month later, in November, they left for Tennessee.

They arrived at Oak Ridge in November and found the lab in some turmoil with uncertainty about its future: the Atomic Energy Commission had taken over responsibility for the Manhattan Project facilities earlier and this included Clinton Laboratories, as Oak Ridge was then called. Most of them were being designated as National Laboratories, promising status and some security, but it seemed until late in 1947 that this might not happen at Oak Ridge. However, within a month of the Russells arriving, the lab was designated as Clinton National Laboratory and then in January it became what it still is, ORNL.

The old Health Division had been split into two in 1946: a Health Physics Division under Karl Z Morgan and a Biology Division headed by Alexander Hollaender. The mission of the Biology Division was to establish an understanding of the health issues that might arise as the AEC drove development of nuclear technology for both civil and military uses – with the impact of weapons as the major concern. It was Hollaender who, on the advice of Muller and Sewall Wright, recruited the Russells. It was recognised that the knowledge of genetic effects arose almost entirely from work on a fly: a biological system quite remote from man. The work that had been done at Bar Harbor suggested that something similar – and much more relevant – could be done with the mouse. It was clear that large numbers of the animals would be required – far more than ever used before – and that the risks of failure would be quite high.

> Although he had no experience of radiation work and none of the kind of industrial scale project that it was clear would be needed, Russell prepared a proposal for the AEC while still at Bar Harbor for an experimental program based on the specific locus test (SLT) – a technique used by Muller on *Drosophila*.[22]

Wild-type male mice would be exposed to radiation and then mated with female mice specially bred to be homozygous at seven loci for a recessive gene with very visible effects. If none of the seven genes were mutated by the radiation then the offspring would appear normal but a mutation at any of them seven loci would result in some offspring being homozygous and taking on the characteristics of their mother. The seven loci chosen included six where homozygosity resulted in distinctive changes to the coat colouring and one that affected the size of the mouse's ears. Russell claimed that even a non-expert would be able to detect these mutations at a rate of 2000 per hour. [23]

Sewall Wright and Muller were both asked by Hollaender to review the proposal and both had problems with it. Wright thought that the there should be a broader look at the impact of radiation, extending over a longer period than the three years proposed by Russell; Muller was concerned that the restriction to seven loci would not give useful results with the number of mice planned. But, after a few concessions the proposal was agreed, the AEC provided funding for three years and Russell started the first task: building up the mouse stock.

The Biology Division had moved into the Y-12 area where, during the Manhattan Project the electromagnetic separation plant had operated as the first stage in separating uranium-235. Space was promised there for the Mouse House but it was slow coming, for reasons of finance and because there was concern that the background radiation level might interfere with the experiment. There was also a risk of contamination from possible accidental releases of activity. So Russell started to build up the special tester stock in temporary accommodation with mouse cages overflowing into the offices. They did not escape all the consequences of the Jackson fire: many of the available mice they needed to set up their new stock went to Bar Harbor where Little was determined to rebuild and re-stock as quickly as possible. Russell managed to get a few animals from a former co-worker but they were of a highly inbred strain and had to be out-crossed to restore some vigour. A suitable cross was eventually found from a Florida professional photographer who bred mice for a hobby and had a garage full of the little creatures. However by early 1949 they were able to move into the promised Mouse House in Y-12 and start a pilot study.[24]

The first radiations were done in March 1949 in a pilot experiment and the first mutation showed up in the eighth litter but they had to wait a "nail-biting" seven months for the second. However by early 1950, with the Mouse House ready, they were able to start the main experiment.[25]

The first published results, in 1951, were from a study using nearly 100,000 mice and 6 Gy exposures of the male mice and examining over 85,000 offspring, he found 53 (or maybe 54) mutations at his seven loci against just two in the control group. Mean $25.0+/-3.7 \times 10^{-6}$ per Gy per locus. This was more than 10 times the rate expected on the basis of *Drosophila* experiments but there was a good deal of variability between loci.[26]

Reviewing the results in Hollaender's book in 1954, Russell concluded that estimates of human hazards based on *Drosophila* might have to be revised.[27]

The difference from *Drosophila* was confirmed at Oak Ridge by Mary

## GENES, FLIES, BOMBS...

Alexander who applied the SLT approach to the little flies in a manner similar to that used for the mice. She found a mutation rate of $1.52 \times 10^{-6}$ per Gy per locus.[28]

In 1958 the Russells revealed a surprise result. Until then it had been received wisdom, based on *Drosophila* experiments, that the efficiency with which radiation produced mutations was independent of the dose rate. The mouse irradiations until then had been at extremely high rates, 0.8-0.9 Gy/minute (probably simulating bomb explosions?) but now the Russells used a much lower rate, a maximum of 0.9 Gy per week. After examining nearly 100,000 offspring of irradiated male mice (and 105,000 from controls), they found that the mutation rate per rontgen fell by a factor of four where spermatogonia had been irradiated.[29] There was no dose rate effect for mature sperm. They put forward a kind of repair mechanism to explain the effect – without specifying what it might be. Of course the factor of four in efficiency had been obtained after a vast change in dose rate – by a factor of some 10,000 – but it was clear that dose rate did have a significant effect. Nine hundred milliGrays was still more than a factor of 100 greater than a worker might expect to receive in a week. It seemed clear that radiation might not be as harmful to humans as previously thought. The doubling dose at the low dose rates was estimated as 1 Gy.

The earlier irradiations reported upon had all been of male mice: the irradiations of female mice with large acute doses reduced their fertility so much that they were made in limited numbers. Studies using about 9000 offspring of irradiated mothers between 1949 and 1951 revealed just one mutation. Clearly it was difficult to draw many conclusions from that but it did have one important implication: the females were unlikely to be more than three times more sensitive than the males and probably less sensitive. As part of the chronic irradiation study some females were irradiated and their 22,000 offspring checked. The conclusion was that the two sexes were similarly sensitive.[30]

Female mouse sensitivity seemed to be a little more complicated than that of males. All the oocytes a female produces are present at birth but remain in an arrested state for much of the time. They complete their development one at a time a few weeks before ovulation. The immature arrested oocyte seemed to be almost completely resistant to to mutation by radiation while the mature oocyte was apparently quite sensitive. This is important because it means that each oocyte is vulnerable for only a short period of its existence. In extrapolating to humans there was concern about a marked difference to mice: immature oocytes in mice are very sensitive to killing by radiation while human oocytes are not. This could have meant that the mutations were not seen because the oocyte had died before maturing. In 1977 Russell showed[31] that this was unlikely: the variations in vulnerability to mutation seen in the mouse were likely to be replicated in humans.

It showed what traps there were in translating results to humans. There was anyway the persistent doubt that the loci studied were not representative because the mice derived from mice discovered by fanciers: they may have been selected just because they had a higher mutation rate giving plenty of interesting variations.

In the UK similar concerns had led the government to invest in research on radiation effects in mammals and they did this through the Medical Research Council. The original plans involved rabbits but the first actual work seems to have been with mice in Edinburgh, where Muller had worked for a while in his travels, in 1947. Here Toby Carter and Mary Lyon from R A Fisher's Department at Cambridge and Rita Phillips from Liverpool started irradiating mice with a plan for experiments based on Russell's SLT method.[32] However it was soon realised that they would need to raise far more mice (it was estimated 150,000 of them) than could be accommodated there and so, after negotiations with the Atomic Energy Authority, a move was made to Harwell in Oxfordshire in 1955. Carter became head of a Genetic Section. Here, in the wake of public and scientific concern about the effects of atomic testing, there were plans for a great expansion of the work and ambitious proposals for the future (involving several million mice) were made by Carter. However, the plans to work closely with the Russells had not worked out well, with the Americans seeming not to release information and not do work they had said they would. This, combined with the obvious fact that the British effort was always likely to be dwarfed by the American one and doubts about whether the funds would ever be available for the UK plans, lead Carter to resign and leave in 1958 for a post at the Poultry Research Centre, where he would become Director. The initial work was published in 1956 but after that the mouse work was left to Lyon and colleagues while Carter seems to have become a world expert on the strength of egg shells and the stresses on them involved with laying.

Mary Lyon and co-workers studied dose rate effects and exposure to neutron irradiation but certainly made a major contribution to the radiation questions with the work on the same stock of mice as the Russells with their methods but with different six different loci. The mutation rates, at $5.0 \times 10^{-6}$ per locus per Gy, were significantly lower than the Americans had found but still three or four times that found with *Drosophila*. It removed the doubts that had arisen that the Russell loci were not representative and more easily mutable because their mice had somehow been previously selected by fanciers[33].

Both Crow and Russell recall a rather rancorous episode of the mid and late 1950s. J B S Haldane submitted a paper to the Bulletin of the Atomic Scientists in which he proposed another method for detecting mutations – and expressed surprise that the Americans had not thought of it. It was based on some complicated calculations the results of which he sent to Crow and Sewall Wright with the suggestions that they could work them out for themselves in just an hour. This they set out to do. It it took several hours – but did result in them having the heady experience of finding an error in the great man's calculations. Haldane's work was published in 1956.[34] In the meantime Muller had waded in, suggesting that Haldane was irked by not being included as an advisor on the Harwell work; Haldane responded by saying that he would no sooner go to Harwell that he would have been to visit concentration camps if invited by the Nazis. In the UK Carter, who was rather committed to the specific locus method, did take the time to look at the practicalities of Haldane's alternative – which involved scoring mutations close to a specific locus – in a pilot study and concluded it was much less efficient than the SLT.[35] Haldane, not surprisingly, disagreed.

The ORNL Mouse House continued to do important radiation research work (not just in genetics but in the effects of radiation on development) well into the 1980s, involving several million mice, by which time it had become a major centre for chemical mutagenesis studies. Bill Russell died in 2003 just shortly after ground was broken for a new facility: the William L. and Liane B. Russell Laboratory for Comparative and Functional Genomics. Budget cuts led to the laboratory closing in 2009 when the remaining mouse stock was moved to the University of North Carolina.[36]

The Harwell mouse work continued into the 21st century although, with the lessening concern about the impact of fallout, the radiation studies declined dramatically. There appears to have been a revival of interest in large-scale mouse experiments to address the threshold question in the 1980s – but nothing came of it.[37] Mary Lyon is best known for her discovery of X-chromosome inactivation, a process which ensures that only one of the two X-chromosomes in females actually functions.

Russell summed up the technical changes in radiation genetics that the mouse experiments had brought at the 1970 International Congress of Radiation Research at Evian in France:

> Estimates of genetic hazards of radiation in man were originally

# RADIATION HARMS

derived mainly from work with *Drosophila*. They are now based largely on the experimental findings obtained with mice. This change in evaluation of risks was not simply a substitution of one set of quantitative values for another, it also involved a major upheaval in what were once thought to be basic principles of radiation genetics. The following had been accepted as important general principles: (1) Gene mutation rate is proportional to radiation dose. (2) Gene mutation rate is independent of dose fractionation. (4) There is no repair of gene mutational damage. (5) There is no threshold dose of radiation below which no genetic damage occurs. (6) There is no recovery from mutation with time after irradiation.[38]

And he continued

The results obtained from work with the mouse have shown that the [above assumptions are] definitely incorrect, and that each one of the six 'general' principles does not apply to mouse spermatogonia and/or oocytes.[39]

## *BEAR AND MRC*

At the time of Castle Bravo a critical point had been reached. The genetic risks of radiation had been confirmed in mice and there was no reason to doubt either that they extended to man or that there was no threshold. The threshold notion still applied to the somatic effects of radiation, both acute and late, but it was weakening and, reflecting this change, "tolerance dose" became "permissible dose". With fallout likely to rise, the consequences of nuclear war more widely understood and the rising interest in nuclear power, there was a need to take stock and say something authoritative and comprehensible to the public. So two committees were set up. In the UK by the Medical Research Council and in the USA by the National Academy of Sciences.

## *THE MRC COMMITTEE*

On 29 March 1955, in the last week of his terms of office, Sir Winston Churchill, in answer to a question in the House of Commons, revealed that the Medical Research Council had been invited to review the existing scientific information on the medical aspects of nuclear radiation and prepare a report that would be published as a Government White Paper. The question was raised by the Labour MP Arthur Henderson[40], later Lord Rowley, who had spoken many times about the dangers of nuclear weapons and fallout, who asked Churchill if the Government would publish a report "on the effects of continuing radioactive contamination of the world's atmosphere."

# GENES, FLIES, BOMBS...

The MRC set up a Committee under the chairmanship of its Secretary, Sir Harold Himsworth[41]. Himsworth, who had done important research on diabetes, presided over the main committee and the two panels it was split into for the majority of the work. One of these looked at the effects of radiation on the health of the individual and the other at possible genetic effects. The somatic panel met nine times; the genetics one eleven times. The main committee, which included Sir John Cockroft, met four times. The activities were supported by two Scientific Secretaries: T C Carter and W M Court Brown. The report, *The Hazards to Man of Nuclear and Allied Radiations*, was published as Cmnd. 9780 in June 1956[42].

The part of the report on the effects on the individual naturally enough focussed on the delayed effects and it dealt first with leukaemia. It drew on data provided by the Atomic Bomb Casualty Commission which showed that between 1947 and 1954 there had been 91 confirmed cases of leukaemia (and 14 suspected) among survivors when only about 25 would have been expected. At Hiroshima, the incidence of the disease quite clearly decreased with distance from the hypocentre. The data to 1954 indicated that the incidence had levelled out in about 1950 but there was no clear indication that it was decreasing with time.

There had been reports of a few cases of leukaemia in patients treated for ankylosing spondolytis. The disease is one which affects the joints, particularly of the spine, and causes severe pain. The joints may then begin to seize up leading to a a loss of flexibility of the spine (a condition known as "poker spine") and restrictions on breathing. One treatment that proved effective was irradiation of the spine (which involved exposure of just about the whole body); it relieved the pain, improved flexibility and sometimes arrested progress of the disease. The Committee sponsored a study of the hospital records of nearly 14,000 patients treated with x-rays between 1935 and 1954. It found 38 cases of leukaemia, about ten times more than would have been expected and there was a clear link between incidence and dose. A quite full summary of the findings of the study, conducted by Court Brown and Richard (later Sir Richard) Doll, was given as an Appendix to the Committee's report and they were later published in full.[43]

The Japanese data and the ankylosing spondolytics study left no doubt that radiation caused leukaemia and indicated that the risk increased with dose. It was recognised that the doses were very large and delivered over a very short time. This kind of acute exposure was not likely to occur as a result of testing or indeed in normal occupational exposure to radiation. However, they could point to the studies of American radiologists obituaries which suggested that chronic radiation could cause leukaemia too.

The Committee considered other effects on the individual: cancers, aplastic anaemia, cataract and reduction in life span were some of them. The effects on the fetus were also reported: abortion, stillbirth and microcephaly.

That radiation could cause cancer could hardly be doubted: the lung cancers in pitchblende miners, the bone cancers of the luminising paint workers and the skin cancers of early x-ray workers all pointed to that. However the Committee in 1956 thought that these were all associated with very large doses which caused severe tissue damage. The threshold principle was still there.

## *GENETIC RISKS AND THE MRC*

The first step in assessing the genetic risks associated with radiation is to estimate the effects of a doubling of the spontaneous mutation rates for various types of disease with a large hereditary component.

As a representative of a dominant trait caused by a single gene they took achondroplasia – reduced stature resulting from disproportionately short limbs. A permanent doubling of the mutation rate would lead to a doubling of the incidence within just three or four generations. Haemophilia is a sex-linked trait where there is reduction in the ability of the blood to clot. A doubling in the mutation rate of the responsible gene would lead to a 90% increase over about six generations and then a slow rise for the final 10% to doubling. Phenylketonuria was taken as the example of a recessive trait. Here the permanent doubling of the mutation rate would cause a very slow increase in incidence of the disease so that it would be expected to rise by 50% only after 50 generations. Incidence would eventually double but only after many more generations.

In order to estimate the overall social impact the report looks at mental deficiency and mental illness which accounted for nearly half the hospital beds in the UK. For severe mental defects a doubling of the mutation rate would lead to an increase of about 3 % in one generation; a permanent doubling would eventually, after very many generations, lead to a doubling of incidence. The three percent corresponded to an extra 1500 cases requiring care.

Records showed that, in 1954, there were about 63,000 cases of schizophrenia and 31,000 cases of manic-depressive reaction under hospital care. These two conditions accounted for about half of all mental illness so an estimate of the effect of doubling the mutation rate would be worthwhile. In the first generation after doubling the incidence of both

# GENES, FLIES, BOMBS...

would increase by around 1%; if the mutation remained doubled then the incidence would increase very slowly, eventually doubling. One percent corresponded to an extra 400 cases of the two diseases.

The report concluded that it was "unlikely that the burden put upon society by a doubling of the mutation rates would exceed by more than a few times the contribution made by the increase of mental disease."

It remained to decide what dose of radiation would actually cause a doubling of the mutation rate.

If all mutations that occur naturally were caused by background radiation hen the doubling dose would be equal to the dose received by the gonads prior to and during the reproductive period of life. Taking this to be the first 30 years measurements of the time gave this as 3 r (30mSv). This would be a lower limit to the doubling dose. However the fraction of mutations due to natural background was thought to be between 2% and 20% suggesting that the doubling dose lay between 150 and 1500 mSv. When the rather sparse data from experiments on plants, insects and mammals (which gave doubling doses ranging from 280-390 mSv) were taken into account the Committee settled on the doubling dose in humans lying between 300-800 mSv.

The Committee did make some recommendations about permissible doses. An individual should not feel unduly concerned about delayed effects (and here they meant leukaemia) provided the dose they received (in addition to natural background) was less than 2 Sv in their lifetime and that it was received over many years. The maximum weekly exposure should be less than 1 mSv averaged over any 13 week period.

As far as genetic effects were concerned they thought that a person could reasonably accept a total dose to the gonads, in addition to natural background, in the first 30 years of life of 500 mSv without concern for his or her offspring. The same restriction could safely be applied to occupationally exposed people provided they did not total more than $1/50^{th}$ of the population. Of course, if a much larger proportion of people might be exposed, the dose would need to be even more restricted to control the social impact of the mutations that would arise. The Committee felt unable, given the inadequacy of knowledge at the time, to give a specific figure for average exposure of the whole population. However they said that they thought it unlikely that any authoritative body would recommend a value more than twice the natural background. In the UK the gonad dose from natural background came out at very close to 1 mSv per year or 30 mSv over thirty years so a recommendation for the UK should be equal or less than 60 mSv.

# RADIATION HARMS

On the question of the risks associated with fallout from weapons tests – which had prompted the Government in the first place – the conclusions were rather reassuring: at the current rate of testing there was negligible hazard. If the rate should rise and more thermonuclear devices be detonated then ill effects might be seen in a small proportion of the population.

## APPENDICES TO REPORT

The report – authoritative, detailed (there were 13 Appendices), rather self-contained and accessible as it was – proved a valuable breakwater as waves of concern and protest about the effects of nuclear testing swept towards Government. Something similar happened in the USA.

## THE BEAR COMMITTEE

The similar action in the USA came with the same aim of producing a statement on the hazards of radiation and testing that might be regarded as authoritative and independent. But it was not a government initiative; the Rockefeller Foundation Trustees agreed to fund a study that would be accepted as mediating between the various current opinions – and be a voice independent of the AEC. Detlev Bronk, the president of the Rockefeller Institute, would commission such a study from the US National Academy of Sciences – of which he also happened to be President. The NAS, it was announced, would make a "dispassionate and objective effort to clarify the issues, which are of great concern and great hope to mankind."

Six committees were set up: on pathological effects of radiation, genetic effects, meteorology, oceanography and fishing, agriculture and food supplies and disposal and dispersal of radioactive wastes. They became known collectively as the Committee on the Biological Effects of Atomic Radiation – BEAR.

The two most interesting committees from our point of view are the one on the pathological effects of radiation, chaired by Shields Warren, and the genetics one chaired by Warren Weaver. The members of the genetics committee were Bentley Glass, Alexander Hollaender and Bill Russell from Oak Ridge, Gioacchino Failla from Columbia, Berwind Kaufmann, Tracy M Sonneborn and Muller from Indiana, Clarence Little, James V Neel, Sewall Wright and James Crow from Wisconsin, Shields Warren, Milislav Demerec from the Carnegie Institution and George Beadle and Alfred Sturtevant from Caltech. All the committees met for the first time in November 1955.[44]

Much of the discussion of the pathological effects group was concerned

## GENES, FLIES, BOMBS...

with the acute effects of radiation. The specific delayed effects discussed were the life-shortening noted among US radiologists and leukaemia. Life shortening could result from exposure of specific organs and consequent development of cancer or leukaemia or from damage to the immune system or to connective tissue or as a result of premature ageing.

In skin cancer and leukaemia there were thought to be threshold effects in operation. After high doses of radiation the blood supply to skin was greatly reduced and there was intractable ulceration and "such chronically damaged skin is a fertile bed for cancer development." Leukaemia "may show manifold increase in persons subjected to a nearly fatal single dose (Hiroshima data) or in those whose professional work has exposed them to higher than acceptable permissible dose rates."

Somatic mutations as a source of cancer are rather dismissed as playing a minor role. The brief discussion of internal sources suggests that the tragic experiences of the radium dial painters will not be repeated with the current standards and that strontium-90 is no real cause for concern. Overall, sticking to the permissible levels arising from genetic considerations will keep the public safe from pathological effects.

The summary report of the genetics committee, written by Warren Weaver, goes to some lengths to stress the grave importance of their work but emphasises that the risks (from radiation) and the benefits (defence, economic and international relations) were not fully understood. In his thoughtful preamble he points out that genetics could answer all the questions that needed to be asked of the science. Although the experts could not necessarily agree on exact numerical values, there was no disagreement on fundamentals. These fundamentals included that radiation (including natural background) causes mutations that are generally deleterious, that the number of mutations is proportional to the total dose of radiation received and that the relevant period to total over is the first 30 years of life. The notion that there was a dose rate (say so much per week) that was safe did not make sense.

However, unity among the experts fell short of agreeing how the damage should be calculated. Some preferred to relate the damage to the number of mutations caused by the radiation and assume that all the mutations cause the same average harm (whether spread over one or many generations); Muller and Sturtevant certainly fell into this camp. Others, notably Sewall Wright, pointed out the difficulties of the genetic load approach.

The two schools of thought are summarised in Weaver's comments:

> One way of thinking about this problem of genetic damage is to

# RADIATION HARMS

assume that all kinds of mutations on the average produce equivalent damage, whether as a drastic effect on one individual who leaves no descendants because of this damage, or a wider effect on many. Under this view, the total damage is measured by the number of mutations induced by a given increase in radiation, this number to be multiplied in one's mind by the average damage from a typical mutation.

Measuring total damage in terms of the number of mutations does indeed necessarily involve this concept of the average damage from a typical mutation, and some geneticists find this concept difficult and illusive. They would point out that mutations may be grouped in classes that differ, on a subjective scale, many thousand-fold in the amount of damage per mutation. As examples they would cite a mutation which results in very early death of an embryo (which might cause very little social or personal distress), and a mutation which results in severe malformation to a surviving child, (which would cause very great personal distress and which clearly involves a social burden).

Sewall Wright would later write in the 1960 BEAR report:

There is one point of view under which the appraisal of genetic damage from increased radiation is a relatively simple matter. If we assume that there is one best genotype and that this is homozygous in all type genes, it follows that all mutational changes from this are injurious and selected against. For each mutation there will be on the average one elimination (or "genetic death") to restore the status quo (in a static population; more than one in a growing population). If we define damage in terms of number of genetic deaths, it follows that all mutations produce equal damage in the long run and it merely becomes necessary to estimate the number of mutations produced by a given amount of radiation to appraise the damage.

There are, however, several considerations that make this point of view unsatisfactory.

In the first place, the concept of a single type genotype probably does not apply to any organism and particularly not to human populations in which extreme diversity is itself essential to a healthy state of society. It is probable that the optimal state of any population is one in which many alleles with slight differential effect are carried at a large proportion of all loci at more or less equal frequencies. Even conspicuously unfavorable effects of mutations in particular combinations may be balanced by favorable effects in others.

In the next place, the equating of all unequivocally injurious mutations is very unrealistic without consideration of the personal and social impact. It will perhaps suffice here to note that the occurrence of a dominant mutation, lethal in the first week of development, will produce no appreciable damage to the population or to any one in it. There will be no appreciable damage to society and little to any person from a mutation that causes a slight reduction in fecundity of otherwise wholly normal carriers in a population that is in balance with its natural resources, and there may be some advantage to a society that is suffering from overpopulation.

On the other hand, a dominant mutation that gives rise to a distressing and incapacitating but not lethal condition that is usually not manifest until after the family is complete may produce enormous personal and social damage before becoming extinct. It is indeed conceivable that there is a class of mutations that endows its carriers with capacities for parasitizing society that cause them to increase in numbers until society collapses.[45]

There was sharp disagreement about how to quantify the genetic harm to the population associated with radiation exposure but the geneticists on the panel did in fact agree fairly well on the number of mutants that would result.

The report introduced the idea of the doubling dose and gave estimates for its value. These ranged between 50 and 1500 mSv with "experienced geneticists" putting it between 300 and 800 mSv. If the population of the US were subjected to a doubling dose in each generation then then rates of genetic disease would eventually double, of the approximately 100 million children born, two million more would suffer genetic defects than would be the case if there was no additional radiation exposure. Perhaps 10% of these would be in the first generation. The doubling would cause "real personal and social distress." If the population were exposed to 10 r rather than a doubling dose there would be about 50,000 new inherited defects in the first generation and this would rise to about 500, 000 per generation if the exposure was continued.

To put matters in some kind of perspective the then-current estimates of the doses to the public received in the first 30 years of life were given. Background radiation over the period averaged out at 43 mSv across the USA with people living at high altitudes receiving as much as 55 mSv.

# RADIATION HARMS

Medical x-rays were adding an average of about 30 mSv. If weapons testing went on at the then-current rate it was estimated they would deliver about 1 mSv annually – although this figure could be in error either way by a factor of five.

Given the closely reasoned argument and the care Weaver took to ensure that they were comprehensible to the interested layman and acceptable to the factions on the committee, the recommendations on permissible levels spring out without clear justification. The first is that no individual should receive a dose to the gonads of more that 500 mSv in the first 30 years of life and the second that members of the general population should receive no more than 100mSv from all controllable sources of radiation[46]. One hundred milliSieverts corresponded to just about two times the natural background in the USA.

The division between the geneticists has been discussed subsequently by James Crow. Another split was between those loyal to the AEC and the disaffected who, of course, included Muller and Sturtevant.[47]

The report was published on the same day (12 June 1956) as the MRC one. After that there would still be differences on how to analyse and express the genetic harm caused by fallout but there could be little doubt that all radiation could cause such harm – there was no safe level and all radiation levels should be kept as low as possible. Ironically, while the report gave some reassurance on fallout, made people aware of the high average levels of radiation from medical x-rays – a subject that Muller had been pushing for nearly thirty years.

All agreed, presumably, on the statement in the overall summary report:

> The inheritance mechanism is by far the most sensitive to radiation of any biological system.

## COLLABORATION

Some historians have treated the appearance of the two reports on the same day as a coincidence but this was far from the case. From the inception of the two studies there had been an exchange of information between the MRC committee and Warren Weaver to ensure that the reports they produced were not obviously in conflict. Much of the ongoing correspondence between Weaver and Himsworth seemed to be related to Weaver's concern that the British approach took insufficient account of societal rather than individual impact. Both Himsworth and Cockroft visited the US.

In the exchange of drafts the two reports approached one another quite closely, in the end differing perhaps more in style than substance: the crisp technical MRC White Paper and the conversational easy read of Weaver's summary.

One choice was the recommendation about the average acceptable exposure of the public. The MRC did not recommend anything specific (Himsworth thought that a political issue rather than a scientific one) but did suggest that whoever did so would be unlikely to choose a level more than twice background. In the UK, this would have been about 60 mSv over 30 years but in the US, conveniently or coincidentally, it would have been 100 mSv – just the value BEAR recommended.

Initially there had been concern that there should be a decent gap between the publications of the two reports to avoid the suggestion of collusion (they were supposed to be independent studies after all) but time pressures narrowed the window within which this could be done. Perhaps, since they were so different in appearance (although so similar in content), it was decided that accusations of collusion could be countered as being simply information exchange – hardly even collaboration – between fellow scientists.

Whatever, Warren Weaver resigned as Chairman of the genetics subcommittee on the day after publication.

## JOINT CONGRESS HEARINGS

In late May and early June 1957 the Joint Committee on Atomic Energy held eight days of public hearings to clarify the scientific issues around fallout and its effects. Leading scientists from around the US were called and made statements. The resulting verbatim reports are a valuable record of thinking at the time on all aspects of fallout. Altogether, the testimony, statements and submitted papers occupy over 2000 pages.[48]

There was unity among geneticists that the no-threshold, proportional principle was soundly based. It was admitted that this has only been demonstrated down to about 250 mSv but several of the witnesses believed that there was much supporting evidence, based on the nature of the damage caused to genetic material, to suggest that it continued down to much lower doses.

On somatic effects there was much less agreement. Most seemed to accept that leukaemia could be caused by radiation but most thought that large doses were required. Shields Warren supported the view of Jacob Furth, President of the American Association for Cancer Research, that there was surely a threshold for leukaemia while H L Friedell, a prominent radiologist with a background in the Manhattan Project and advising government, took a somewhat hesitant position saying that he "would

# RADIATION HARMS

hesitate to accept this concept that a threshold does not exist." Ernest Pollard, chair of biophysics at Yale, believed, on theoretical grounds, that proportionality was plausible and should be supported as the conservative choice. The geneticists, when asked, suggested that somatic effects like leukaemia and life-shortening did not have a threshold believing that their basic cause was closely related to genetic effects.

As far as induction of tumours was concerned Friedell pointed out that they were produced in animals but only at very high doses. Warren said that that there was no evidence that tumours were more frequent in adult populations exposed to total body radiation.

Asked about the genetic damage caused by fallout, Crow said that the current testing rate would eventually lead to, at most, a 3% increase in mutation rate and probably much less. Sturtevant, later, complained that many geneticists had been concerned about the AEC statement that testing had caused no damage to people saying that even tiny percentage increases corresponded to tens or hundreds of thousands of people suffering around the world. Muller, appearing after Sturtevant, made the same point and upped the numbers involved by a factor of ten, saying that hundreds of thousands or millions of lives would be seriously curtailed or damaged. He also complained about suppression of material in the late 1940s and about his treatment by the AEC over the Geneva paper.

## *INTERNAL DOSE*

One of the problems that dated from the early days of radium was that associated with radioactivity that entered the body. It had been seen as a source of harm for radium dial workers, for pitchblende miners and, with the generation of all kinds of radio-toxic materials in the atomic weapons programme, for workers and, through fallout, for the public. It was a serious problem for two reasons: once radioactivity enters the body it may stay there and irradiate the organs for a very long time and, while it is there it may be in very intimate contact with sensitive organs and tissues.

The residence time in the body depends on the chemical nature of the radioactive material and its the radioactive half-life. At worst the material can have a very long residence time and end up incorporated into a sensitive tissue. Radium for example finds its way into the bones and more or less stays there, irradiating away, for the rest of your life. Radioactive iodine ends up in the thyroid gland, irradiating that until its radioactivity declines away. This is not quite at worst because there is something special about alpha-particle emitters – alpha-particles, energy for energy being the most deadly of the bunch.

Because they ionise densely, the source of their enhanced hazard, they

# GENES, FLIES, BOMBS...

have a very short range. While a beta-particle can be stopped by a sheet of tin and an x-ray can pass through people, alpha-particles can be stopped by a flimsy sheet of paper. So, provided the radioactivity is outside the body, alpha-particle emitters pose little hazard. Even close to the skin their effect is reduced by the layer of dead cells that covers our body.

Once the material is inside the body it's a different matter. To take a popular example: plutonium when ingested hurries off to the bones and sits there for a lifetime, its alpha particles irradiating the growing bone surfaces. One of the major concerns of the time was not plutonium (well not for the public anyway) but a radioactive isotope of strontium present in fallout.

Radioactivity is measured in units of Becquerel with 1 Bq being one atomic disintegration per second. The Becquerel replaced an earlier much larger unit based on the disintegration of a gram of radium called the Curie with 1 Curie being equal to 37 billion Bq. All the radioactive quantities have here been converted into Becquerels.

## *RADIOSTRONTIUM AND SUNSHINE*

Concern grew about strontium-90 because it was known to be an abundant product of fission. It is chemically similar to calcium, an essential element in the body, and would therefore pass through the food chain and concentrate in the bones. The permissible level of the time for workers was that the bone should not receive more than 3 mSv (0.3 rem) in a week and it was possible to calculate that this corresponded to a total of about 37 kBq (1μcurie) in the whole body – about 37 Bq per gram of calcium.[49] The measurements which had been made and reported in the MRC report had all been less that one thousandth of this.

In the USA the AEC had set up two secret research projects on the spread of strontium-90. The first of these GABRIEL, a review of existing available data on the hazards of fallout by Nicholas Smith at Oak Ridge, concluded that by far the most significant isotope was Sr-90. The cheerily-named project SUNSHINE was set up in 1953 to gather data on fallout levels worldwide in biological products and systems. The products included cheese; the systems included bones from human cadavers, notably children and still-born babies. SUNSHINE's results were considered reassuring by the AEC but the project remained secret because of concerns that released data could give valuable information to the Soviets. Also, since the body parts (perhaps several thousand) were being gathered from around the world without parental permission or at best on false grounds, there was some nervousness that the supply could dry up if the project went public. There was a limited release of information in

# RADIATION HARMS

1956 which was indeed rather reassuring in tone.

One of the outcomes of the project was perhaps the most inappropriately-named scientific unit of all time: the Sunshine Unit (SU). It corresponded to 37 mBq of Sr-90 per gram of calcium in the body. It was rather quickly renamed the Strontium Unit.

## TELLER AND PAULING

On 20 February 1958 there was a debate between Linus Pauling and Edward Teller on the television station KQED-TV in San Francisco. Sparked by Pauling's January petition opposed to nuclear weapons, signed by more than 9,000 American scientists, it had the title *Fallout and Disarmament* and the subtitle T*he nuclear bomb tests...Is fallout overrated?* The highly structured debate lasted an hour.

It went along predictable lines. Both men wanted peace; but not, said Teller, if the price was loss of freedom. Both would welcome disarmament; but Teller would not trust the Russians. It was a polite reflection of the myriad debates going on around the world at the time.

On the specifics about the effects of fallout they deployed arguments that were familiar at the time. Pauling said that there were 75 million babies born around the world and 2% of them – 1.5 million – were defective. He had estimated that 1% of these defects were caused by radiation from fallout at the current level – and, he said, George Beadle and James Crow agreed with him. This meant that 15,000 children were born defective annually directly as a result of bomb testing. And since one large bomb could produce as much fallout as a whole year of testing, the man who authorised the testing of such a bomb would doom 15,000 children to some kind of severe disablement.

Teller disputed the figure saying there was no real evidence that radiation in small doses was harmful. He had calculated the consequences of fallout as a 0.1% rise in mutations and so 1500 defective children world wide – but this was a measure of our ignorance rather than a genuine estimate. He thought it very unlikely that any genetic defects would occur at all. If there were mutations, well they were the fuel of evolution.

## LIFESPAN REDUCTION

An effect of radiation that was discussed by BEAR, the MRC and in the Joint Hearings was the reduction in lifespan. As early as 1944 this had been seen in rats and it was later confirmed in mice. It was put down by some to an acceleration of the normal ageing process perhaps through a degradation of the immune system. The doses used were large – a significant fraction of the lethal dose – and the effect was nowhere near as

# GENES, FLIES, BOMBS...

clear (and usually absent) at lower dose rates. There were even experiments that showed that doses of 1 mGy per day actually increased the lifespan of mice – although making them more prone to leukaemia. There was no evidence for the effect in humans until Shields Warren published in 1956 an analysis which showed that US radiologists died a few years younger than physicians generally. It did not take long for someone to point out that the differences was due to different age structures of the different parts of the medical profession not the effects of radiation. [50]

By the 1960s there was a general consensus that life shortening at lower doses was due to the induction of leukaemia and solid cancers rather than to non-specific damage.

### *POPULAR CULTURE – CINEMA*

Many people learned about radiation from the cinema, magazines and newspapers.

After the second world war there was an explosion in science fiction films and short features made by Hollywood; between 1948 and 1962 nearly 500 were made when there had been just a handful before that. But, even from the earliest days of commercial cinema, radiation had played a part.

The silent *The Invisible Ray* of 1920 seems to be lost but the 1936 version remains. In this renegade scientist Janos Rukh mounts an expedition to darkest Africa to study an ancient meteorite which he believes to be a source of radium. Rukh manages to find the meteorite but is exposed to it and is saved from death by radiation poisoning only by an antidote provided by a colleague. However, from this time anyone he touches instantly dies. To made things worse, the antidote has the particularly unfortunate side-effect of inducing paranoid rages, and this coincides with Rukh's wife falling in love with another man. He also believes that other members of the expedition have stolen his discovery. Things do not end well as he seeks general revenge using the gift of death given him by radiation.

In *The Thing from Another World* in 1951 the eponymous star is an intelligent but radioactive vegetable-based humanoid. It is the pilot of a flying saucer that crashes at the North Pole. After surviving a destructive attempt at recovery by a small US task force despatched to investigate, it thaws out and snacks by killing and absorbing the blood of sledge dogs in this Howards Hawks film from 1951. It proves impervious to gunfire – bullets are not the best way to kill a carrot – but succumbs to an improvised electric arc. The radioactivity appears to be incidental.

## RADIATION HARMS

However, *Attack of the Crab Monsters* of 1957 is a different matter. In this, after an opening sequence of exploding H-bombs, a team arrives on an uncharted island close to Bikini atoll to look for an earlier expedition which has disappeared without trace. The expedition had been sent to investigate the effects of radiation on the sea and plant life. It soon emerges that the island has been taken over by giant, intelligent, talking mutant crabs – their intelligence having been enhanced by eating the brains of the previous visitors. The crustaceans (which resemble *Macrocheira kaempferi*) set about consuming the brains of the new arrivals in an environment that constantly rumbles and shakes because of the crabs submarine activities. The island is slipping into the sea.

The three survivors find themselves trapped on a headland with a mutant crab the size of a small bungalow creeping towards them waving long pincers. Grenades, guns and a small hand axe seem unable to deter the thing so one of the three climbs a nearby radio mast and brings it down electrocuting the crab and himself too. The remaining survivors – who happen to be a handsome American and a petite brunette in a tight sweater – cling together. Is it over? Who knows?

Writer Roger Griffiths later recalled that he had been told to cram every scene with horror or suspense. This, in a budget-limited sort of way he did but recalled that "when I went to go see it, the audience fell asleep!"

*Attack* had clearly picked up on the notion that radiation may cause mutations; other films concentrated on its somatic effects.

*The Amazing Colossal Man* of 1957 features US Army officer Lt Col Glenn Manning who suffers massive radiation burns from a plutonium bomb blast. These heal completely within a day (although his hair falls out) but he starts to grow into a 60-foot giant. The strain on his heart and the trauma that must be associated with spectacular expansion make him go mad and he rampages through Las Vegas clad in no more than a tennis-court-sized loincloth. An attempt to reverse the growth with an inoculation, using a syringe the size of a man, goes dreadfully wrong when he spears a doctor with the needle and kills him. He picks up his girlfriend in his huge hand and lumbers onto nearby Boulder Dam. After he responds to cries of "Put the girl down!" by putting her down, he is shot by the Army and plummets into the Colorado River.

In the sequel the following year, *War of the Colossal Beast,* Manning is found alive in the Mexican desert and is returned to Los Angeles for memory restoration. This fails and he goes on another orgy of destruction before coming to his senses and committing suicide by clutching some electrical power lines. This time, audiences may have thought, as they

## GENES, FLIES, BOMBS...

dozed, lets hope it's final.

The variety of effects that could be caused by radiation is illustrated by *The Incredible Shrinking Man*. In this 1957 film Rober Scott Carey is exposed to a mysterious shimmering radioactive cloud while out boating and then to insecticide when he passes a truck spraying trees. The combined effect of these agents is to make Carey shrink while retaining perfectly his proportions. An antidote temporarily arrests contraction with Carey just one metre tall. He forms a relationship with a female dwarf and appears in a sideshow but the antidote wears off. He returns home and soon he is just a few centimetres tall. He survives an attack in the basement of his home by the family cat and, as he reaches the millimetre range, one by a spider. Soon he is small enough to escape through the wire mesh window screen when he becomes reconciled to his fate which seems to be to reach atomic dimensions.

The film anticipates the synergistic (and possibly multiplicative) effects of radiation and chemicals – later found in the combination of radon exposure and smoking.

The genre includes *Spiderman* (first seen in a comic book in 1962), who acquired the legendary arachnid characteristics of strength and agility after a bite from a radioactive spider. In *Behemoth the Sea Monster* of 1959 radioactive waste dumping in the ocean disturbs a gigantic creature that can project bolts of lightning and radioactive beams. A rampage through London is brought to an end only when a radium torpedo is fired into the creature's mouth.

However the epic that sums it up and brings us back to the theme of genetic heritable effects – is the Oscar-nominated (for its special effects) *Them!*

In this 1954 Warner Brothers film mysterious and violent deaths in the desert near Alamogordo in New Mexico, the site of the Trinity A-bomb test in 1945, bring the FBI. Trailers have been ripped open, stores demolished. There are giant footprints and a victim is found pumped full of formic acid. Myrmecologists are called in and soon they encounter a three metre long ant. They identify it as a mutant caused by radiation from the atomic bomb test.

The nest of the insects is destroyed but not before two mutant queens have escaped to found new mutant colonies. One is discovered in a ship and destroyed by naval gunfire. The second queen sets up home in the storm drains under Los Angeles. Her nascent nest is destroyed by the Army using flame-throwers.

# RADIATION HARMS

The film closes with the senior biologist warning:

> When man enters the atomic age, he opens the door to a new world. What we may eventually find in that new world, no-one can predict.

With round about one billion cinema tickets being sold in the UK each year in the 1950s and 50% of the US population going to the movies each week, this popular genre, largely nonsense, must have had some impact on popular notions of the dangers of radiation.

We can add to this post-apocalyptic films. In the low-budget *Five* of 1951 five people survive a nuclear holocaust because they are in safe places at the time. One lady is in a lead-lined x-ray room; a man is in the elevator in the Empire State Building. Soon, after murders and disappearances, there are just two facing an uncertain future.

*Unknown World* of 1951 is rather more fantasy. To escape a nuclear war a party burrows into the earth in a "Cyclotram" (atomic-powered) and enters a vast underground cavern complete with subterranean sea. But it is bleak news that the rabbits that they have taken with them as food have been made sterile. A volcano erupts. Some escape to a tropical island and an uncertain future. More serious is *3000 A.D.* (*Captive Women* in the USA) where, post-apocalypse, the "Norms" and the "Mutates" fight to win the devastated city until they join forces to combat the invading "Upriver People". In *The Day the World Ended* of 1955 the radioactive fallout is just one problem. The other is a monster which kills anything it comes across and, if it is contaminated, eats it, him or her. However it proves to be soluble in water so rain kills it and it seems that the precipitation might wash the fallout contamination away. The two survivors (handsome man and toothsome woman) who have escaped the radiation and the creature stroll into "a new beginning."

*The Day the Earth Caught Fire* of 1961 has a different theme. Weapons tests cause the earth's orbit to change so it moves closer to the sun. Things get hot. The only way to save the planet is to detonate more weapons and flip the earth back. Did it work? We can't be sure. There is a grimmer, grittier and much more serious genre too represented by Nevil Shute's novel *On The Beach* made into a film in 1957.

The radiation/mutation genre reached its peak in the late 50s with some ten films a year but declined to just one or two per annum after that[51] Coincidentally or otherwise mirroring atmospheric testing.

### *POPULAR CULTURE - PRINT*

*Mechanix Illustrated* was a popular magazine which started in 1928

# GENES, FLIES, BOMBS...

Its December 1953 edition featured an article by Otto Binder[52]. It illustrates how fairly good science could be translated into something readable and sensational.

The title is *How Nuclear Radiation Can Change Our Race*. The main illustrations show a ragged-trousered bunch of terrified nuclear war survivors being hosed with radiation and illustrate the two possible outcomes. The first of these, labelled "FRIEND" shows a slim bald man teaching a science class of earnest people. He holds forward a test-tube of liquid and gestures in a way that suggests he is imparting wondrous knowledge. He has a large bald domed head and is wearing an outfit with a very short skirt. A larger illustration is titled "FOE". In this, dome-headed men again reign but now they have guns, evil twisted sneers and are clearly intent on slavery or extermination. They are wearing rather alarming glossy shorts.

In the text the "cold hard facts" that show that atomic radiation has changed us physically are examined:

> There have been 154 cases of cataract at Hiroshima and Nagasaki.

> Hermann J Muller says that in the next 1000 years there will be as many "genetic deaths" in Japan as there were casualties from the bombs; genetic damage will haunt Japan for the next 30 generations.

> Even the smallest amount of radiation is harmful says Karl Z Morgan – although there is a danger level of 0.3 r below which repair takes place.

> X-rays can stunt children's growth and damage the reproductive organs.

> It can cause anaemia.

Nothing too silly there. And there is even some reassurance: test fallout does not pose any significant hazard.

But nuclear war will create changes in genes and 400 million babies may be affected in the next 1000 years. T H Morgan's *Drosophila* experiments (Binder calls the flies, rather confusingly, "winged guinea pigs") produced mutants without wings, with extra legs and two heads. The "even greater Mutant Maker" Muller had created an incredible number of completely new species that bred true using x-rays. What would be the effect of atomic war and the gamma-rays produced on human genes.?

It could result in "throwback forms of inferior mind which would then unleash a horde of new 'sub-men' upon civilisation." On the other hand a

# RADIATION HARMS

superior species might be born "Homo Superior". This could be a worse result than the Homo Inferior alternative: Superior could wipe us out. Or maybe just enslave us.

Binder even suggests what Homo Superior might look like. Perhaps 6½ feet tall he would have a broader head and a bigger brain (perhaps 50% bigger) but be hairless. Thinner finer bones, smaller and fewer teeth and (an interesting detail) no useless wisdom teeth. Altogether slimmer he would lack the "recessive" fifth toe. Superior in intelligence, this Mutant Man might become a wise, scientific leadership class taking mankind to a future of reason, peace and brotherhood with an end to prejudice and intolerance.

Among the solid adverts for Print Your Own presses, Heavy Duty D-C Arc Welders, Snap-on Tools and 52 Piece Socket Sets it all, no doubt, sounded pretty plausible to readers.

It would be a pity not to mention the Miss Atomic pin-ups.[53]

Las Vegas, determined to capture the kudos of being the atomic bomb capital of the USA, promoted a series of photographs of showgirls dressed in swimming costumes with a bomb theme. "Miss Atomic Blast" of 1952 and "Miss A-Bomb" of 1953 appear to have slipped into oblivion but photographs still circulate of "Mis-Cue" being crowned with a mushroom cloud in 1955 and of showgirl Lee A Merlin celebrating the 1957 Plumbbob series of tests, a mushroom cloud covering the front of her swimwear. Mis-Cue marked a further delay in the many-times-postponed operation Cue.

## BOMB STUDIES 1950s-1977 – SOMATIC

A Japanese physician, Gensaku Obo, is generally credited with finding the first tumours that could be associated with radiation in 1956 but it was Harada and Ichida who looked at the data from the Hiroshima tumour registry and found a definite link. They published their results in an ABCC Report in 1959 and it was subsequently reprinted in a journal.[54]

They found a excess of malignant neoplasms for the period May 1957 to December 1958 that fell with increasing distance from the hypocentre, paralleling the incidence of leukaemia. The 24 cancers of the stomach registered from people within 1500m of the hypocentre (the point directly below where the bomb had detonated) were twice what was expected and the 10 lung cancers were four times. The increases were significant at the 1% level. Cervical and ovarian cancer were increased at the 5% level.[55]

The decline with distance from the hypocentre at Hiroshima was confirmed, although the excesses were less, when the timespan for the group was extended to March 1960 and a rather different population sample used by Zeldis. A Nagasaki population showed no significant effects (it included far fewer cancers) – or rather confusing ones since the non-exposed population appeared to suffer a higher incidence of malignancy than those who had been exposed.

There were apparent inconsistencies. In 1962 for example Beebe reported on the causes of death of mortality of a sub-sample of 20,000 in the LSS, including exposed and non-exposed survivors and saw excess leukaemias but no other effects of exposure.[56]

However by its 1964 report UNSCEAR[57] could point to the excess of malignancies nearly equalling the excess of leukaemias to date.[58] The excess was rising and, given the apparently longer latent period for malignancies, this could be expected to continue.

ICRP6 in 1964 mentions explicitly leukaemia as a somatic effect. Other serious effects were possible:

> Late somatic injuries include leukaemia and other malignant diseases, impaired fertility, cataracts and shortening of life.[59]

In discussing the possibility of the existence of a threshold for leukaemia the ICRP pointed out the difficulty in establishing what this was (based on the US radiologists data) and noted that:

> The most conservative approach would be to assume that there is no threshold and no recovery, in which case even low accumulated doses would induce leukaemia in some susceptible individuals, and the incidence might be proportional to the accumulated dose. The same situation exists with respect to the induction of bone tumors by bone-seeking radioactive substances. (p18)

While excess incidence and mortality were seen, it was not possible to relate this to dose until the doses that had been received by the survivors were reliably estimated. Some early figures came in 1957 (referred to as T57D) but it was not really until 1965 that a set of data (T65D) that won the confidence of workers became available.[60] Using them, a detailed study of the location of survivors at the times of the explosions made it possible to estimate the doses they actually received taking account of shielding.

By 1972 it was clear that leukaemias were slowly returning towards more normal levels: the number of deaths dropped from nearly six times the expected level in 1950-1954 to about two times. However, the deaths from malignancies began to rise, nearly doubling in the late 1960s. It meant that these cancers were starting to emerge some 20-25 years after the exposures and there was no way of knowing when they would start to decline.[61]

With the T65D dose estimates it was possible to calculate excess risk as a function of dose received – rather than of just distance from the hypocentre. The plots of these confirmed – although with considerable uncertainty – that risk rose with dose received. They were generally consistent with a linear, no-threshold, relationship. It began to look as if the excess absolute risks per unit dose from several cancers were already comparable or even greater than those from leukaemia. And of course these malignancies were only just beginning to show.

One of the other advantages of knowing the doses was that the Japanese bomb data could be compared with that from other sources. So UNSCEAR could compare the leukaemia risks with those deduced from the ankylosing spondolytics study, with results for the lung from tuberculosis patients and with thyroid cancer risks extracted from infants irradiated in the neck region to treat enlargement of the thymus (a condition that proved of no consequence). There seemed to be a good degree of consistency.[62]

The 1977 UNSCEAR report had a few more years of data. The main concern, given continuing excess deaths, was to estimate what the lifelong mortality risk might be – to anticipate how long the excess might continue. They relied on a deduction from the US radiologists and ankylosing spondolytics that, in those instances, eventually, the deaths from solid tumours had been a few times those from leukaemia. The numbers of cases this conclusion was based on were fairly small but they settled on a factor of five for the ratio of the total number of eventual fatal malignancies to those from leukaemia. It led to a total predicted lifetime risk of dying from cancer as a result of whole body radiation exposure to $200 \times 10^{-4}$ per Gy. This was based on exposures of more than 1 Gy and they felt able to reduce it by a factor of two for the lower dose rates that would be occur in routine exposures.

The situation in Japan in 1978 was reviewed by Kato and Schull[63]. Of the 80,000 people in the extended LSS 23,500 had died by 1978, nearly 5,000 from some form of cancer. One hundred and eighty had died of leukaemia and of these about 90 were reckoned to have done so as a result of radiation from the bombs. Of the 4576 who had died from other forms of cancer it was thought that about 160 deaths could be attributed to radiation from the bombs. Of these the most common cancers were stomach (42 deaths), colon(16), other digestive (24), lung(32) and breast(15). The leukaemia deaths had returned, more or less, to their background level; the other cancers had clearly not even peaked, with a dramatic jump in risk apparent in the 1975-1978 period.

## BOMB STUDIES 1950s-1977 – GENETIC

Early studies aimed at finding any chromosomal damage suffered by survivors were unsuccessful. The procedure adopted – a painful needle biopsy from the testes – gave chromosomes but these were difficult to study. There was even some doubt about the number of human chromosomes. Better techniques were developed and by the mid-1950s damage was being seen. These cytogenetic studies of survivors continued to show effects and it was established that chromosomal damage increased with estimated dose. The techniques developed were used in radiation accident dosimetry. It was a different matter in the study of possible transmissible mutations.

While the initial Genetic Programme, with its clinical aspects, was terminated in 1954, the collection of data on mortality of children of parents who had been in Hiroshima or Nagasaki at the time of the bombs (grouped according to whether the parents had been within 2000m of the hypocentres or beyond) was continued as the F1 Mortality Study. This included a control group of children whose parents had not been in the cities. This was based on residency, birth and death information already collected by the Japanese authorities. The first results were published in 1966[64] and were based on about 50,000 births registered between mid-1946 and the end of 1958. The study, in the authors' words, "failed to disclose significant variation in mortality ascribable to differences in radiation." A follow-up study of the cohort published in 1974[65] reached similar conclusions now based on mortality of the children up to an average age of 17 years.

By 1973 the US was unwilling to continue funding the ABCC and was seeking a cost-sharing arrangement with Japan and this led to the setting up, in 1975, of the Radiation Effects Research Foundation (RERF), a jointly managed and funded research organisation.

Although the clinical and mortality studies of offspring had shown no significant effect of parental exposure up to the end of the programme in the 1970s, the knowledge that radiation produces mutations in all other species tested led to a continuation of the programme in the form of extension of the cytogenetic (chromosomal) studies begun in 1967. These did show that survivors had suffered significant and persistent chromosome damage clearly related to the doses they had experienced. Analysis of the chromosomes of in leukocytes in blood samples taken from 2885 offspring of survivors had, by 1975, failed to show any significant effects compared to a control group of just over 1000 children.[66]

There were other sources of information about the effects of radiation. Some of them have been mentioned. But the bomb survivors came to completely dominate the search for an understanding of the risks of radiation for the rest of the century.

## ICRP26 – CHANGE IN EMPHASIS?

In 1977 ICRP issued Publication 26, a thoroughgoing revision of their recommendations which recognised that the main hazards in most likely situations were the stochastic ones of cancer and hereditary effects. Their severity was not related to dose but their likelihood was. This likelihood was proportional to dose. Thresholds disappeared. Any radiation, however small the dose, carried with it a risk of causing a fatal cancer or a serious hereditary disease. But the lower the dose, the lower the risk.

The approach to control adopted was risk-based and, with the assumption of proportionality between dose and risk, brought together several elements in a neat and unified system. For example, the risks associated with non-uniform irradiation of the body could be estimated relatively easily – and that meant dealing with internal emitters was much simpler. It also brought together cancer and hereditary effects.

The risk estimates for cancer were largely deduced from the Japanese bomb data and the risk of a fatal cancer from uniform irradiation of the whole body was considered to be about 0.01 for each Sievert. The hereditary effects relied on data from experiments on small animals and lower organisms for the effects combined with observations of the background levels of hereditary disease in man. The detriment associated with hereditary effects was considered to be less that than from the somatic ones. The serious hereditary effects in the first two generations were likely to be about 0.004 per Sv and about twice this overall

## LATER DEVELOPMENTS

More cases of cancer that could be attributed to the bombs continued to be found, including new cancers. There was no indication that a peak had been reached. As the data accumulated it became possible to study how the risks to survivors depended on their age at the time of exposure and how they changed as they aged. The was still a need to extrapolate the data into the future if the final impact of the bombs was to be estimated and there were various models proposed to do this[67]. There were marked differences between cancers and ideas changed as more data became available but by the end of the century some discernible patterns emerged that made prediction of the final number of cancer deaths from the bombs to be possible. The two patterns of greatest significance were that the risks of developing cancer were generally higher if the survivor was younger at

# GENES, FLIES, BOMBS...

the time of exposure and the excess relative risk (the cancer risk from irradiation as a fraction of the natural cancer risk without irradiation) declined as survivors aged. The latter factor was confused by the fact that most "natural" cancer rates increase significantly as we get older.

Both these trends were suggesting in the early years of the 21$^{st}$ century that the peak of attributable deaths was some way off. People who were younger at the time of the bomb were forming a larger and larger part of the surviving cohort. More significantly, while the ERR appeared to be declining as people aged, the background cancer rate increased with age, accelerating ever faster. The absolute risk associated with radiation was thus increasing too. In 2003 Preston and his co-workers[68] concluded that the overall effect was that the death rate from cancer resulting from the bombs would continue to rise until peaking in the year 2020.

The patterns were confirmed in 2012 with the follow-up to 2003[69]. By 2003 58% of the LSS cohort of 86611 survivors, had died, about 11000 of them from solid cancers. About 5% of these deaths, 527, were attributed to the bomb. To this had to be added the about 100 deaths from leukaemia that had been caused by the bombs in the early years.

Cancer was of course not the only somatic effect. Apart from the vast numbers who died from early radiation effects there were whatever abortions that were induced in women pregnant at the time of the bombs, the brain developmental problems and later effects in children who had been irradiated in the womb. It also became apparent that there were other somatic effects manifest later in life in survivors which included increased rates of several disease categories. Osaza estimated the excess in the cohort at 353 up to 2003 – an increase of about 1% over the expected number.

Overall the late excess deaths in the cohort added to nearly 1000 and it seemed that they would be climbing for another twenty years. The situation with heritable effects in the early years of the 21$^{st}$ century was very different. Still, none had been found.[70]

## BASELINE RATES OF GENETIC DISEASE

Until 1993 the estimates of baseline line frequencies used by UNSCEAR were based on the work of C O Carter in the 1970s. In summary Carter estimated that the total incidence was 125 per 10,000 live births: 95 autosomal dominant, 25 autosomal recessive and 5 X-linked. After a review the Committee very nearly doubled the incidence rates to a total of 240 per 10,000 live births, now split among the three categories 150, 75 and 15. The reasons for the changes are several but broadly they are a combination of the inclusion of new diseases, better data on the ones already in Carter's analysis and a conservatism that reflects a better understanding of how much more there was still to find out about genes and genetic disease.

## DOUBLING DOSE

Estimates used up to 2000[71] were based, with one exception, on mouse data for spontaneous and radiation-induced rates and there was a consensus that for chronic irradiation the doubling dose lay near 1 Gy.

In 1999 Neel could see an argument for a significantly higher DD – as high as 5 Sv.[72] It was a reminder of how things had changed since the heated days of the fallout debate, being nearly 200 times more that the 0.03 Sv suggested by Haldane in 1955 in a spat with John Cockcroft.[73]

There was a substantial revision in thinking in 2000 about how the DD should be calculated when it was recognised that there were substantial differences between spontaneous mutation processes in mice and man. The preferred method of estimation of DD became the use of human spontaneous rates (rather than those of the mouse) and mouse radiation-induced rates. While this was a significant conceptual change it led to a DD close enough to 1 Gy for this value to be retained for risk estimation.[74]

## MUTATION COMPONENT

When the mutation rate is increased from its background level the rate of associated disease will increase until it reaches a new equilibrium where the increased mutation rate is balanced by the failures to reproduce because of the disease.

For a dominant mutation, if the mutation rate is permanently doubled, it means that the disease rate will be doubled too. How long it takes to reach the new equilibrium depends on the selection coefficient – the likelihood that an individual with the disease will fail to reproduce. The lower the selection coefficient, the longer it will take to reach the new equilibrium.

If the increase in mutation rate is for one generation only the disease rate will rise in the next generation and then fall back to the original level; the lower the selection coefficient, the longer the return will take. However, whatever the selection coefficient, the total extra cases of the disease will be the same: Muller's genetic death idea.

The relation between disease incidence and mutation rate is expressed through something called the Mutation Component (MC).

*Fractional increase in disease rate=Mutation Component x Fractional increase in mutation rate*

# GENES, FLIES, BOMBS...

For a dominant mutation, the Mutation Component becomes 1 in just a few generations quite quickly.

For purely recessive mutations the MC reaches its final value of unity much more slowly. It takes more than 100 generations for it to rise significantly; it is taken to be zero in the first and second generations. For an incompletely recessive gene the Mutation Component behaves in a fashion between dominant and recessive over the generation with the interesting (although probably not really significant) characteristic that it could eventually rise to a value somewhat greater than unity.[75]

## THE MISSING MUTANTS – PRCF

There was not much confidence that any inherited effects would be found in the children of the bomb survivors when the genetic studies were set up and none were. This eventually led to an examination of the relationship between the data from the mouse experiments and the real impact on humans.

The Potential Recovery Correction Factor (PRCF) – an impenetrable name – was a factor which bridged the gap between mouse and humans[76].

Essentially the mouse experiments – and all the others on animals that had been used to establish the efficacy of radiation in producing mutations – had required that the mutations studied, while causing visible changes in the heterozygote phenotype, did not significantly affect the possibility of a live birth of a heterozygote. If they, did there would be little to count. They had, in the jargon, to be "recoverable." There could be many mutations that were lost because the damage caused by radiation resulted in some failure in development of the fertilised egg. Sankaranarayanan pointed out that the main effect of radiation interaction with DNA was deletion – of a part of a gene or, quite often, of a whole genomic region. So a mutation of a particular gene could fail to be passed on because an intact gene was essential for survival or because the collateral damage caused to nearby genes by the radiation event meant that there would be no live birth. The latter was more likely in gene-rich parts of the genome.[77]

> A total of 63 human genes (in which spontaneous mutations result in autosomal dominant and x-linked diseases) was included in the analysis. The results show that with the criteria used, only in 21 or in one-third of the genes, induced mutations may be compatible with viability and hence potentially recoverable in live births.[78]

When normalised to account for the structure of the genome it was concluded that only between 15 and 30% of mutations caused by radiation would be recoverable: the PRCF is between 0.15 and 0.3. It meant that the doubling dose was effectively between 3.3 and about 6 Sv – quite compatible with the Japanese experience.

So were reconciled – provisionally perhaps – the results of the mouse experiments of half a century earlier with the conclusions of fifty years of analysis of the genetic consequences of the bombs.

*RISKS – MENDELIAN GENETIC*

|  | First generation offspring | First and second generations |
|---|---|---|
| Autosomal dominant and X-linked | 750-1500 | 1300-2500 |
| Autosomal recessive | 0 | 0 |
| Cases per million live births for continual exposure to 1 Gy per generation. | | |

So the risk estimates for Mendelian genetic disease at round about the end of the century, taking all these changes into account and with a DD of 1 Gy, were, summarised by UNSCEAR in 2001, as shown in the table.[79]

It is difficult not to look back to the debates over the first BEAR report and Muller's genetic death idea. Muller himself would likely have been outraged by the solution to the problem of genetic deaths in future generations: simply ignore those beyond the second. To take recessive conditions out of the equation (quite literally) in this way – without any explicit justification – would have seemed unethical and been intolerable. The PRCF idea would probably have seemed more interesting and much more acceptable: it used detailed knowledge acquired about the genome to elaborate one of the ideas that made Sewall Wright question the genetic death approach. Not all genetic deaths are the same; some would pass unnoticed as the development of a fertilised egg stopped at an early stage. Not completely acceptable though; the equivalence between such an early and unknown filtering out of a non-viable bundle of cells and a later spontaneous abortion or still-birth implied in the PRCF approach would surely have triggered an argument. Muller would have wanted that sorted out.

By the end of the century the inconsistencies in the understanding of Mendelian diseases seemed to be on the way to resolution – although with

# GENES, FLIES, BOMBS...

null data on humans there remained the possibility of some surprises – but a whole new area had opened up when people started to address multifactorial diseases. The Mendelian forest proved hard enough to navigate; multifactorial diseases were a jungle in comparison.

## CONGENITAL ABNORMALITIES

Congenital abnormalities are those present at birth (the term implies nothing about cause) and, in varying severities, occur in a few percent of live births (estimates put frequencies at between 1 and 7% across the world). A small proportion appear to be inherited in a Mendelian fashion but most are multifactorial with a genetic and environmental component. The characteristics are that someone biological related to a sufferer is more likely to suffer the same abnormality than a member of the general population is and the closer the degree of the relationship, the more likely that is. Identical twins, for example, are more likely (although not certain) to suffer the same abnormality than dizygotic ones.

The genetic component seems also to be associated with several genes. In some cases, of course, there may be no genetic component at all and the abnormalities are associated with environmental factors such as prenatal infections or known teratogens. These are the characteristics of multifactorial diseases.

## ADULT DISEASES

Many diseases of adulthood also have a multifactorial character: diabetes, coronary heart disease and essential hypertension are common examples. In all of these, genetic factors predispose people towards developing a condition but whether they do or not depends on, for example, lifestyle. Some diseases are associated with mutations in just one gene; in others the susceptibility may arise from a mutation in several different ones. Some diseases require mutations in multiple genes before susceptibility to them is increased significantly.

## RISKS OF MFDs

From the 1960s there have been many detailed and complicated models to explain the evolution and maintenance of polygenic variation by a balance between mutation and selection. Falconer , in 1965[80] introduced the Multifactorial Threshold Model (MTM) and a parameter the disease "liability"; if this exceeded a threshold then the individual would develop the disease. The liability had two components: a genetic one and one that reflected environmental factors. The genetic one involved many genes that might be mutated and affect the disease liability and it might reasonably be taken, because of the central limit theorem, to be normally distributed.

# RADIATION HARMS

The environmental component distribution would also be normal. The MTM then asserted that if the sum of the genetic and environmental liabilities exceeded a certain threshold level the the disease would develop.[81]

While many of the parameters in the model might be unknown, something that could be calculated was the heritability (conventionally $h^2$) of the disease. This could be done by examining the incidence of the disease in relatives of sufferers and remembering that individual genes were passed on in a Mendelian fashion[82]. The heritability of common MFDs generally lay between 0.3 and 0.9. Using the MTM it was possible to estimate bounds for the heritability of the associated liability.

In order to make a risk estimate for radiation protection what was needed was a knowledge of how a change in the mutation rate would affect the disease frequency. This is the origin of our old friend the Mutation Component – and in this context the name makes sense.

To understand this the Finite Locus Threshold Model (FLTM) was developed [83]

This took an example of just five genes – five loci – each of which could be turned off by a mutation. The liability that came from these was combined with another random variable representing the environmental factor to give the total liability. When this was greater than the threshold value the person was deemed to develop the disease and to have a selection coefficient of s. Offspring inherited individual genes in a Mendelian fashion.

ICRP used the model in its Publication 83[84].

The situation was too complicated to solve analytically and so simulations were run on a computer, tracing the disease incidence through the generations. A range of values representative of the MFDs was taken for the parameters and the Mutation Component calculated for each one.

The conclusion of the calculations was perhaps surprising: for the range of heritabilities of interest the Mutation Component was about 2% in the early generations. This was the number used in the risk estimation.

## RISK ESTIMATION

The MC is about 0.02 for each early generation. What remains is to decide the PRCF for multifactorial diseases. If the average chance of recovering a mutation in one locus is PRCF1 then the chance of a mutation being recovered simultaneously at n loci is $PRCF1^n$. The PRCF1 estimate

# GENES, FLIES, BOMBS...

for autosomal dominants is in the range 0.15-0.3 so, if there were just two loci involved the PRCF would be between $0.15^2$ and $0.30^2$: between 0.02 and 0.09. There are probably more than two loci so this would be an overestimate and it was the conservative value range used by UNSCEAR.

The natural occurrence of multifactorial diseases is very high; UNSCEAR took 65% as representative. The risk per Gray in the first generation is then, with a doubling dose of 1 Gy, in the range 250-1200 cases per million. Developmental abnormalities were assessed (by a rather different argument) at the higher rate of 2000 cases per million per Gray in the first generation. For the second generation the chronic disease risk in the same as the first (the MC is the same); the risk of abnormalities is somewhat higher than in the first.

## SUMMARY OF RISK ESTIMATE CHANGES

The balance of concern about the effects of radiation have changed quite dramatically over the second half of the century. Before the Second World War the threshold notion meant that, provided dose were kept under control, the risk of cancer was considered very near zero. By the 1970s, with the analysis of effects on the Japanese bomb survivors, it was realised that cancer was a possible significant outcome. Just how large the risks were assessed to be depended on the model used to project into the future. The additive model was favoured by UNSCEAR and ICRP in the 1970s and 80s and this suggested a fatal cancer risk of around 0.01 per Sv for low dose rate exposures. As more data were gathered and analysed the risk numbers grew, not least because the favoured model now became the multiplicative one. By 1990 the risk was widely believed to be around 0.05 per Sv and this continued to be the accepted value into the 21st century.[85]

The assessment of hereditary risks went the other way. In the 1970s the risks were seen as being towards 0.01 per Sv, in the 80s onwards around 0.002 per Sv emerged as a popular figure for first generation risks.

So, while the risks of genetic disorders were seen to have fallen by a factor of about five; the cancer risks had gone up by a similar factor. The precise factors of change are arguable but even (disputable) estimates make it clear what a radical change there has been in the last 40 or so years. Early in the 21st century, in its 2007 Publication 103, the ICRP estimated the detriment – the total harm – from hereditary effects of radiation as about 4% of that from cancer.[86]

# 10     A BETTER SPECIES?

> Democratic control, therefore, implies an upgrading of the people in general in both their intellectual and social faculties, together with a maintenance or, preferably, an improvement in their bodily condition.[1]

H J Muller

*MULLER – POLITICS*

Here is little doubt that, up to his experiences in Russia, Muller was a left-wing radical. In his support of *The Spark* and later in his eugenics address (*Dominance of Economics*...)[2], with use of terms like "class war", the tone is one of rather extreme views even within the left-wing spectrum. It seems likely, given his enthusiastic deployment of Marxist and Communist rhetoric in, for example, his letter to Stalin that he was more than just this. He was, by the general understanding of the word, a Communist. He may not have been a member of the Party, as he reassured the President to get the job at Indiana but, for a while at least, he swallowed the philosophy.

He left Russia because of the oppression that came with Lysenkoism, the purges that were going on and, ironically, Stalin's alarming reaction to his book *Out of the Night* – as we will see shortly. The irony was that the book was later quoted as an example of Muller's Communist sympathies and affiliations.

We know that the FBI kept a file on him for many years as a result of his radical student associations and *The Spark* episode. The Censor kept a close eye on his correspondence during the war as evidenced by a letter to him (addressed to "Mr Henry Muller" at Amhurst) requiring explanation of a telegram from Lancelot Hogben in Birmingham, England in 1943. The cable[3] read:

"ARISTALESS DEAD WANTS ARISTALESS AND ARISTAPEDIA BADLY LUV"

The Censor was justifiably mystified by the references to *Drosophila* mutants.

The FBI visited him at Bloomington in early March 1953 to serve a subpoena requiring him to appear before the House Committee on Un-

## GENES, FLIES, BOMBS...

American Activities (HUAC) in Washington later in the month. Muller had remarked before being told this that he was in a good mood because he had just learnt that Stalin had died; now he was concerned that the appearance before the committee would interfere with his plan to deliver a lecture on the suppression of science in the USSR.

The subpoena could not have been a complete surprise. The Committee, under its new Chairman Harold H Velle, had announced that it planned to investigate the penetration of American education by Communists. Bloomington had set up a committee to deal with the question of the loyalty oath that administrators and academics were being required to sign. By the end of March the 36-strong Association of American Universities had announced that membership of the Communist Party "extinguishes the right to a university position."

Muller went to the university President Herman B Wells to warn him and Wells reassured him that the university would stand by him and suggested he talk to their lawyer.[4] There was really no time for this and on 14 March 1953 he made his appearance before a closed session of the HUAC. There were elements of farce at the start.

The first question asked of him was whether he knew an Alfred Kantorowicz. This clearly referred to the Communist International agent of that name who had been in the USA from 1941 to 1946 before being deported. Muller replied that he did know an Alfred Kantorowicz but that was his father-in-law, Thea's father, a retired and rather distinguished professor of dentistry. However he did not know the Comintern agent. The confusion of the two men was, he said, a mistake made by the Nazis too.

He had been asked to appear simply as a result of this confusion and was free to go. However the Committee showed an interest in his views on why Communism seemed so attractive to intelligent people and he stayed for some time to answer that question and others[5]. It was an amicable and beneficial outcome for Muller and Wells could write in his memoirs[6] that:

> "...Dr Muller made a brilliant appearance before the committee and created such a favorable impression that the complimented him on his views. He returned home in triumph..."

*The Daily Banner*, an Indiana newspaper, gave him a headline[7]: "Muller Opposes Reds in School" and reported that he "opposed Communists teaching science, psychology, psychiatry and a host of other fields." It would be "disastrous".

Perhaps the HUAC appearance marked a change in official attitudes to

# A BETTER SPECIES?

Muller – although some, such as Willard Libby of the AEC, never came to trust him. For most though he was no longer the suspected Communist in the long shadow of his earlier pronouncements and dalliances with Russia. Maybe he had not quite admitted everything, said he was wrong and apologised but he was ready to condemn the Soviet philosophy and their actions. Never a cold-war warrior and still a gad-fly nagging about radiation standards and arguing for peace. But, tellingly, while he warned of the dangers of fallout, he saw the associated risks as smaller that those of Soviet domination.

By most standards he became a liberal. He did campaign for a bilateral weapons test ban, he did warn of the dangers of radiation (as he had always done) but he devoted some time to another life-long interest: eugenics.

## *EARLY EUGENICS*

The domestication of animals is thought to have started with the dog and it seems to have been between 30,000 and 15,000 BC. The first livestock to be domesticated were sheep and maybe goats around 10,000 BC followed, within perhaps a thousand years, by cattle and pigs. Cats, those favourites of the Pharaohs, were made pets about the same time. Different domestications started in different (and probably multiple) places. Many began in The Fertile Crescent that extended from the eastern Mediterranean through what are now Turkey, Iraq and Iran. But domestic sheep may well have originated further east, dogs may have been tamed first in Western Europe and it seems likely that the horse was first harnessed and ridden on the Russian Steppes. Wherever it all started, it has been going on for a long time.

Some animals would have been more suited to their owners' needs than others: dogs that were tamer or better hunters, horses that were stronger with more stamina and pigs that were plumper. It needed no knowledge of genetics to see that it might be worth breeding from the better (we should maybe call them "more suitable") animals. So selective breeding began and it may have produced quite quickly recognisable improvements in stock. As agriculture and commerce developed the animals would have been traded and the knowledge spread.

Successful selective breeding for particular traits needs careful recording – of the pedigree of mated animals – and vigilance to ensure that other unfavourable traits are not being bred in. It is something of a balancing act. Written breeding records are known to have been kept by Arab horse breeders from the 14th century and oral records were probably passed down long before that. By medieval times horses were being bred in

## GENES, FLIES, BOMBS...

Europe for size, stamina, speed, agility and temperament for war and everyday use.

Much credit is usually given to Robert Bakewell (1725-1795) of Dishley in Leicestershire, England as the father of scientific animal breeding. This may be a rather Anglo-centric view of things but it is clear that Bakewell achieved some remarkable results with horses and cattle through the employment of "in and in" breeding: a process in which animals were mated with close relatives to achieve the required traits. This resulted in rapid development of selected traits but, of course, ran the risk of exposing unwelcome genes, resulting in deterioration, inbreeding depression and infertility. Bakewell's major achievement may well have been achieving improvements in the selected traits while managing the unwelcome ones. He also appears to have been canny in the selection of the stock he started with. His notable successes were improved Longhorn cattle and the New Leicester or Dishley sheep, bred for speed of growth and quality of meat.

With all this animal breeding going on right from antiquity it would have been incredible if no-one had considered the potential for improvements in the human stock through management of breeding. While there is a long history of preventing supposedly defective people from mating it was generally by simply killing them and the reasons were not just (if at all) eugenic.

The Spartans, in the 1st millennium BC, were famously fearsome warriors of classical Greece with a social system directed entirely toward military ends. The Spartates, those privileged to become warriors, passed through a rigorous training programme, the *agoge*. But there was an early hurdle that had to be leapt: newborn boys were subjected to physical inspection soon after birth and, if they had any perceived defects, they were left out on a hillside for a few days. If they survived that they might be accepted as potential Spartates. Some versions of the story have defective babies being simply thrown into a gorge and left to die. Whether this was just the disposal of the supposedly worthless burdens on society or whether it was thought that it would preserve the Spartan stock (and hence be truly eugenic) is not clear; no-one (wisely no doubt) quizzed the Spartans too closely on the matter. It seems likely that these warriors had little interest in animal breeding so might not have extrapolated the insights of that to themselves.

Plato, a little later, was more philosophical and seems to have made the connection:

> And how can marriages be made most beneficial? – that is a

## A BETTER SPECIES?

> question which I put to you, because I see in your house dogs for hunting, and of the nobler sort of birds not a few. Now, I beseech you, do tell me, have you ever attended to their pairing and breeding?
>
> In what particulars?
>
> Why, in the first place, although they are all of a good sort, are not some better than others?
>
> True.
>
> And do you breed from them all indifferently, or do you take care to breed from the best only?
>
> From the best.
>
> And do you take the oldest or the youngest, or only those of ripe age?
>
> I choose only those of ripe age.
>
> And if care was not taken in the breeding, your dogs and birds would greatly deteriorate?
>
> Certainly.
>
> And the same of horses and animals in general?
>
> Undoubtedly.
>
> Good heavens! my dear friend, I said, what consummate skill will our rulers need if the same principle holds of the human species.

And slightly further into the text:

> Why, I said, the principle has been already laid down that the best of either sex should be united with the best as often, and the inferior with the inferior, as seldom as possible; and that they should rear the offspring of the one sort of union, but not of the other, if the flock is to be maintained in first-rate condition. Now these goings on must be a secret which the rulers only know, or there will be a further danger of our herd, as the guardians may be termed, breaking out into rebellion.

(*Plato's Republic* Book V 360 BC, Jowett translations, Gutenberg 2008)

Plato's solution was essentially a ballot, rigged to ensure that the right people married and mated. Positive eugenics, you might think, at its most inventive best. Sadly it was coupled with negative provisions too[8]:

# GENES, FLIES, BOMBS...

> ...the offspring of the inferior, or of the better when they chance to be deformed, will be put away in some mysterious, unknown place, as they should be.

Infanticide was legal in ancient Rome:

> Table IV: Rights of fathers over the family Law 1 "A notably deformed child shall be killed immediately."
>
> *The Twelve Tables of Rome* of about 450BC (taken from Yale Law School website)

> We put down mad dogs; we kill the wild, untamed ox; we use the knife on sick sheep to stop their infecting the flock; we destroy abnormal offspring at birth; children, too, if they are born weak or deformed, we drown. Yet this is not the work of anger, but of reason – to separate the sound from the worthless.
>
> Seneca *Moral and Political Essays: On Anger* Book I, 41-51AD

It was practised, in the context of short and brutal lives that hovered around subsistence, in many other societies. But perhaps with no thought of eugenics, rather of practicality and even compassion.

Moving on to more recent times, an essay from 1913 may well have the distinctly racist flavour of its day but it may also be not far short of the truth:

> Infanticide and Exposure, terms which in early ages were virtually synonymous, appear on first consideration to have been practised among uncivilized tribes for a bewildering multiplicity of reasons. (1 McLennan, "Studies in Ancient History," chap. vii., passim.) There is the female infanticide of China and the Isles of the Southern Pacific, the male infanticide of the Abipones of Paraguay, and the indiscriminate massacre of the Gagas, who, killing every child alike, steal from a neighbouring tribe. There are the Indians who offer up children to Moloch or drown them in the Ganges; the Carthaginians sacrifice them to Kronos, the Mexicans to the rain god. There is the murder of twins and albinos in Arebo, and the cannibalism of the Aborigines. In Mingrelia, "when they have not the wherewithal to maintain them, they hold it a piece of charity to murder infants new born." There are the Biluchi, who kill all their natural children, and there is the modern factor of shame.
>
> Co-existing with all these various practices there is the definitely Eugenic motive. Among the Aborigines, all deformed children are killed as soon as born. The savages of Guiana kill any child that is

# A BETTER SPECIES?

"deformed, feeble, or bothersome." The Fans kill all sickly children. In Central America "it is suspected that infant murder is responsible for the rarity of the deformed." In Tonquin we hear of a law which forbids the exposing or strangling of children, be they ever so deformed. In Japan, deformed children were killed or reared according to the father's pleasure. Among the Prussians the aged and infirm, the sick and deformed, were unhesitatingly put to death.[9]

JBS Haldane could, in the 1960s, tell a story of positive eugenics

> My friend G. C. Dash informs me that until recently the Jats, in northern India, along with ordinary fraternal polyandry, practiced eugenics as follows. A young man judged of out-standing merit for physique, courage, and other good qualities, was allowed access to all married women of a village. He was given a pair of gilded shoes which he left outside the door when performing his eugenic duties, to warn off any ordinary husband. After fifteen years or so, when his daughters became nubile, he was killed to avoid inbreeding. But he might, and often did, leave the village with 'a chosen partner. Having fought in the same brigade as the 6th Jats, I can testify to their courage and efficiency as soldiers. In view of such traditions, the choice of a father other than a woman's legal husband may arouse less opposition in some parts of India than in other countries, whether artificial insemination or the normal process is employed.[10]

The gilded shoes give the story its particular genteel, exotic quality but there was, of course, not always a happy ending for their owner. Which leads to perhaps the best known example of eugenics through selective breeding: that practised in the Oneida Community, so-called because they established their headquarters on forty acres of upstate New York.

## ONEIDA COMMUNITY – STIRPICULTURE

The community's founder John Humphrey Noyes, a Yale theologian and preacher who shortly after receiving his license to preach announced that, after long study (and possibly some kind of mental breakdown), it had been revealed to him that the perfection demanded of man by God could be achieved on earth – God would not expect the impossible. Indeed the Millennium had already arrived (in 70AD) and sinners and saints had been separated since then. It led him to claim (presumably since it was his idea) to be morally perfect and quite incapable of sinning. These unconventional beliefs lost him his license to preach but, with the evident confidence that such revelations usually bring, he promptly established a "Perfectionist" church in New Haven CT. In 1840 he set up a community of around 30 friends in Putney, VA where all things, including marriage partners, were held in common. It was when his neighbours objected – Perfection was one thing, free love another – that the community relocated to Oneida in 1848.

By now two doctrines were established: complex marriage and male continence. Complex marriage codified the fact that sex was permitted with any other consenting adult of the opposite sex – subject to the mediation of a female elder – and male continence meant refraining from ejaculation during or even after sexual intercourse – it was considered wasteful. There were arrangements for training young men in the art – the duty fell to post-menopausal women; young women were required to undergo initiation by older men. In an 1853 article[11] Noyes wrote that the community was opposed to "involuntary procreation" and looked forward to "scientific combination" when "amativeness can have its proper gratification without drawing after it procreation, as a necessary sequence." Thus might the race be raised from the ruin predicted by physiologists.

His principle was in place from the early days of the Community: James Humphrey himself fathered four children with four different women between 1841 and 1858. However, in 1868 he introduced what he called "stirpiculture"[12] and set out his ideas in *Scientific Propagation* in 1870[13]. This was a much more formalised system that used in-and-in breeding to create a breeding stock and then careful mating slightly more widely to avoid the risks this brought long term. It was "..an attempt to create a new race by selecting a new Adam and Eve, and separating them and their progeny from all previous races."(p108)

Out of the 300 strong community 81 men and women became parents of 58 children, not with their spouse but with a selected other member of the community. The procreative unions were all approved in advance by senior members of the community on the basis of spiritual, intellectual, moral and physical qualities of the proposed matings. Amative sex was permitted provided the male continence technique was employed. The children of the unions were looked after in infancy by their mother but soon after they could walk they moved into the "Children's House" and were cared for communally – with visits from parents. Most of the fathers involved in the experiment had only one child. John Humphrey fathered a further nine (with nine different young women) – bringing his total to thirteen. He was presumably helped by a principle he stated in his 1870 article: "Providence frequently allows very superior men to be also very

# A BETTER SPECIES?

attractive to women, and very licentious."

The pedigrees of the authors of a 1923 book[14] give some indication of the complexity of the Oneida family tree. One author, Hilda Herrick Noyes, was a grandchild of John Humphrey Noyes's sister Charlotte Augusta and she had a child with one of John Humphrey's sons, John Humphrey II. Hilda's father was one James Barton Herrick but Hilda's mother, Tirzah Crawford Miller Noyes, had another child with her uncle, John Humphrey I's brother George Washington Noyes, and this was the co-author George Wallingford Noyes.

The Community was troubled when Noyes tried to hand over the leadership to his eldest son Theodore Richard – a professed atheist. The experiment was finally abandoned in 1879 after John Humphrey slipped away on a June night, under threat of a rape charge, across the border into Canada from where he advised the winding up of the complex marriage system and stirpiculture. The community then set itself up as a Joint Stock Company with more usual personal relationships and concentrated (they always had business interests to support the Community) on the manufacture of cutlery – the company "Oneida" became an international home-ware brand and still trades today. Anita McGee was right when she wrote in 1891 – "Stirpiculture was planned to insure the future of the church and the community: stirpiculture destroyed both."[15]

In his 1870 *Scientific Propagation* Noyes refers to both Darwin and Francis Galton. Darwin is quoted at length – from his 1868 *The Variation of Animals and Plants under Domestication* – to show what is possible from selective breeding. Francis Galton, coiner of the word "eugenics" and often seen as its father is called "a late English writer" by Noyes and is accused of "meekest conservatism" – because he thought so much could be achieved just by wise marriages.

## *GALTON*

We have met Galton already as the founder of biometrics but a reminder of his wide interests comes from the banner of the galton.org website:

> Victorian polymath: geographer, meteorologist, tropical explorer, founder of differential psychology, inventor of fingerprint identification, pioneer of statistical correlation and regression, convinced hereditarian, eugenicist, proto-geneticist, half-cousin of Charles Darwin and best-selling author.

He took up the eugenic theme at the end of the first part of *Hereditary Talent and Genius*[16] in 1865 where, having established that talent and genius are inherited, he conjured up a Utopia in which examinations had been developed which tested every important quality of mind and body "and where a considerable sum was yearly allotted to the endowment of such marriages as promised to yield children who would grow into eminent servants of the State." He had in mind ten young men and ten young women; each couple would receive a wedding present of £5000 and their children would be maintained and educated for free.

It would address the problem of society as he saw it. Life was much more complex than it used to be but man had not advanced to cope with it, even less to take advantage of it: "We are living in a sort of intellectual anarchy, for the want of master minds."

He repeats the theme in the second part of the paper[17] but that is more concerned with the characteristics of races – and the general superiority of the Anglo-Saxons. We learn that the American Indian is "naturally cold, melancholic, patient and taciturn" while the Negro "has strong impulsive passions, and neither patience, reticence nor dignity" – although he is loving towards his master's children. We learn that a "large part of an Englishmen's life is devoted to others, or to the furtherance of general ideas, and not to directly personal ends." North Americans (presumably the non-indigenous ones) are "enterprising, defiant and touchy; impatient of authority; furious politicians; very tolerant of fraud and violence; possessing much high and generous spirit, and some true religious feeling, but strongly addicted to cant."

The characteristics are all, through inheritance, consequences of the history of the races and mainly the action of various kinds of selection – natural and social. So a eugenic agenda lies, generally, just beneath the surface in both parts – clearly apparent really only in the elite marriage idea.

Things are much the same in his book *Hereditary Genius*[18]. Although there is much more detail, data and arguments about the transmission of genius, Galton only puts forward timely and judicious marriage as a solution. What he did was to attack the Church vigorously for the eugenic (although he did not yet have the word) damage it has done:

> Thus, as she [the Church]—to repeat my expression—brutalized human nature by her system of celibacy applied to the gentle, she demoralised it by her system of persecution of the intelligent, the sincere, and the free. It is enough to make the blood boil to think of the blind folly that has caused the foremost nations of struggling humanity to be the heirs of such hateful ancestry that has so bred our instincts as to keep them in an unnecessarily long-continued

## A BETTER SPECIES?

antagonism with the essential requirements of a steadily advancing civilization. p358

By the early 1870s his enthusiasm for judicious marriage had waned, largely because he realised that so little was really known about heredity that any eugenic programme might do more harm then good. So he spent the best part of the next 20 years investigating that. His main tool was statistics and his studies hardly helped: they seemed to suggest that, for any real change in physical and mental constitution it would be necessary for selective breeding to go on forever. Should it stop, the race would regress to its old ways. His rabbit experiments to test Darwin's pangenesis theory led him to his stirp theory of inheritance but this had its problems. In his preface to the 1892 edition of *Hereditary Genius* he acknowledged that racial betterment had "hardly advanced beyond the stage of academic interest."

In 1883 he introduced the term "eugenics" in his *Inquiries into Human Faculty*...[19] in a footnote on p17:

> That is [he is referring to the term "eugenic questions"], with questions bearing on what is termed in Greek, *eugenes* namely, good in stock, hereditarily endowed with noble qualities. This, and the allied words, *eugeneia*, etc., are equally applicable to men, brutes, and plants. We greatly want a brief word to express the science of improving stock, which is by no means confined to questions of judicious mating, but which, especially in the case of man, takes cognisance of all influences that tend in however remote a degree to give to the more suitable races or strains of blood a better chance of prevailing speedily over the less suitable than they otherwise would have had. The word *eugenics* would sufficiently express the idea; it is at least a neater word and a more generalised one than *viriculture* which I once ventured to use.

By 1901 there was much more data and Pearson had formulated a modified theory of stirp that suggested that changes from selective breeding might be more permanent than previously thought. In October Galton delivered the Huxley Lecture at the Anthropological Institute under the title *The Possible Improvement of the Human Breed under the existing Conditions of Law and Sentiment*.[20] While this did draw on much wider data it came up with no new formulation beyond the judicious marriages of the elite of 30 years before – other than a suggestion that the wealthy and the noble should take promising young people under their wings. This "befriending" would guide and support them in marriage and child rearing. He called this "the augmentation of favoured stock".

# GENES, FLIES, BOMBS...

Much of the argument, dressed up with statistical distributions, is based on the idea of civic worth of the likely offsprings of couples. Quite how that is to be evaluated is not detailed – although Galton suggests educational attainment and peer evaluation for the elite categories. He quotes – seemingly approvingly – the work of a Dr Farr who estimated the worth of the baby of an Essex labourer as £5 and that of an elite baby as several thousand.

If there was any doubt of the need for eugenics Galton suggested that you only had to look to Britain's imperial role:

> To no nation is a high human breed more necessary than to our own, for we plant our stock all over the world and lay the foundation of the dispositions and capacities of future millions of the human race.[21]

Galton's enthusiasm for judicious marriage remained until the end; he had no other concrete proposals. Perhaps he didn't need them: he was part of a self-confident, well-heeled society at its imperialist peak and still glowing with the achievements of the industrial revolution. There was the hard-to-prove idea that civilizations had fallen because the fertility of their leaders had fallen but this was just theory (and the Queen, just dead, had managed nine children with a man of high civic worth). No hint of apocalypse there.

However, others saw a much darker future:

The national fibre was unravelling. The poor quality of the recruits for the Boer Wars, where almost 80% of volunteers from Manchester had to be rejected because they were physically unfit, put the future defence of the Empire in doubt. The systematic studies of the poor of London sponsored by the shipping magnate Charles Booth over the period 1899-1902 had shown that the levels of deprivation were much higher than anyone expected: 35% of Londoners were living in abject poverty. The most powerful and enduring results of the study were the maps of the city detailing in colour just where the poor lived. Swathes of the capital of the world were marked in black ("Lowest class. Vicious, semi-criminal"), dark blue ("Very poor, casual, Chronic want") and blue ("Poor"). Seebohm Rowntree, in his 1901 *Poverty, A Study in Town Life,* showed that things were similarly grim in the city of York.

The weak, sickly and inadequate (the feeble-minded) were no longer removed efficiently by natural selection because they were protected by over-generous welfare provisions and they produced weak, sick and inadequate children. They were a growing burden on society. The Royal Commission of 1908[22] estimated that nearly 1% of the population were

# A BETTER SPECIES?

mentally defective.

The degenerate (criminals, deviants, alcoholics and the feckless and irresponsible poor ) were not just a burden but also a threat: data showed that they were more fertile than the upper levels of society.

A more general concern was the differential birth rate. The birth rate had been declining since the 1870s, something of a worry for eugenicists in itself, but more alarming still was that the decline was not uniform across society. The lower classes were out-producing the higher ones. The 1913 Birth Rate Commission found that in the upper middle class there were 119 births /1000 married males under 55. The corresponding figures for the lower classes were higher: for skilled workmen 153/1000 and for the unskilled 213/1000. The death rate among the poor was higher but not by enough to compensate. Karl Pearson put it as 25% population producing 50% of the next generation. The best stocks were dying out; the worst were increasing. It was a recipe for national degeneration.

Two Government reports were commissioned in a relatively short space of time:

*The Interdepartmental Committee on Physical Deterioration 1903*

This was set up to examine the widespread view that the lower classes were in a declining physical state (based mainly on the problems of army recruitment). The evidence they collected suggested otherwise; public health was getting better rather than worse. They recommended banning the sale of tobacco to children, publicising the evils of drink and a more rigorous approach to monitoring the health of children.

*The Royal Commission on the Care and Control of the Feeble Minded (1904-1908)*

This recommended that keeping mentally defective men and women from becoming parents would reduce significantly the number of mental defectives in the population. It specifically excluded surgical or other artificial measure to achieve this as they were unlikely to be publicly acceptable.

## LSE 1904 DEBATE

To advance the eugenics cause a meeting was arranged by the Sociological Society at the London School of Economics in May 1904[23]. Galton spoke briefly – nothing too specific but a statement of the importance of eugenics and of the need to educate people in the principles of heredity. He did mention the importance of judicious marriage.

The discussion showed that this meeting at least was not convinced

# GENES, FLIES, BOMBS...

about eugenics. Karl Pearson was generally flattering personally to Galton but Dr Maudsley had doubts that the rules of heredity were well-enough established to support a eugenics programme; Dr Mercier thought they were so obscure and complicated that they might as well be just chance; Dr Warner pointed out that the very best families gave rise to degenerates. Several speakers spoke of the importance of the environment and the need to improve it before getting too zealous about eugenics.; Benjamin Kidd suggested that the problems of the lower races were not inherited intellectual deficiencies but social ones. H G Wells threw in the thought that he was:

> ... inclined to believe that a large proportion of our present-day criminals are the brightest and boldest members of families living under impossible conditions, and that in many desirable qualities the average criminal is above the average of the law-abiding poor and probably of the average respectable person. Many eminent criminals appear to me to be persons superior in many respects – in intelligence, initiative, originality – to the average judge.

He asserted that progress would come not through planned marriages but only from eliminating the weak.

George Bernard Shaw put it apocalyptically:

> ...there is now no reasonable excuse for refusing to face the fact that nothing but a eugenic religion can save our civilization from the fate that has overtaken all previous civilizations.

His remedy was an acceptance of promiscuity: "What we need is freedom for people who have never seen each other before, and never intend to see one another again, to produce children under certain definite public conditions, without loss of honor."

Galton was clearly not best pleased. He had appeared with the Tablets and not been made welcome as the great prophet[24]:

> When this debate began, I was extremely unhappy at the quality of it. The two first speakers [presumably Maudesley and Mercier] really seemed to me to be living forty years ago; they displayed so little knowledge of what has been done since. More than one of the later speakers were really not acquainted with the facts, and they ought not to have spoken at all. We are much indebted to Professor Weldon for raising the debate to a higher level.
>
> ...
>
> Dr. Hutchison believes that environment is far more important than

# A BETTER SPECIES?

> stock, but you know perfectly well how one baby, dog, horse, differs enormously from another by nature; and surely it is not denied that we should take pains to increase the multiplication of the best variants.
>
> ...
>
> As to Mr. Kidd, I do not attach importance to his points. His drones would have selected the best drones, and each one would have selected the best of its kind and worked out their own civilization in their own way.
>
> I have little more to say, except that I do feel that if the society is to do any good work in this direction, it must attack it in a much better way than the majority of speakers seem to have done tonight.

The hereditarian principle subscribed to by eugenicists was that inadequacy, degeneracy, deviance, fecklessness and a propensity to poverty – like talent, genius, musical ability etc – were all largely inherited characteristics.

In two articles in the *Times* in 1906 (11/10 and 16/10) Sidney Webb, the socialist, set out arguments that the birth rates among the best class of person were falling while those of the degenerate classes were rising. His remedy was the positive eugenic step of rewarding and supporting childbearing by the worthy and providing scholarships for the children.

## THE POOR LAW REPORT

The British Poor Law system stretched back to Tudor times and earlier. The last major revision, in 1834, took the existing system of parish workhouses and formed them into larger Unions with bigger establishments. Paupers were to live in them and receive relief at a starvation level – to discourage freeloaders, criminals and cheats.

> The default assumption was that the cause of destitution was the moral fault of the individual: a "failing of character". Poverty was generally seen as a voluntary condition, with the pauper not so much the victim as the perpetrator of his own distress. Deterrence and punishment were therefore to be central features of Poor Law relief.[25]

The system began to break down at the end of the 19th century partly because of the scandals that emerged: exploitation of children and women and extreme deprivation (in the scandal at Andover workhouse the inmates were routinely sucking the putrid marrow from animal bones they were crushing for fertilizer.) Piecemeal changes meant that it was seen as not just confused but as inefficient and expensive: the Poor Law was costing 50% more per pauper in 1905 than it had ten years before.

The Royal Commission on the Poor Laws and Relief of Distress was set up as one of its last acts by the Conservative Government of Arthur Balfour in August 1905. The formidably knowledgeable Commission – which included Charles Booth and Beatrice Webb – spent four years gathering massive amounts of information but was ultimately split on what needed to be done. They issued a Majority and a Minority Report. Broadly the Minority report – supported by Webb and the other Fabians – proposed a more radical change than the Majority one. But the division of the Commission meant that the Liberal government in place at the time of the reports in 1909 felt obliged to do very little. So the principal interest lies in the Minority report, largely written by the Webbs and re-printed in large numbers as a Fabian tract, as a milestone in the development of thinking about welfare. From our point of view it gives some insight into progressive ideas on eugenics at the time.

In comments in *Eugenics Review* Sidney Webb[26] first criticised the Poor Laws as anti-eugenic largely because the mixed Workhouses encouraged immoral behaviour among the degenerate and and wanton procreation among the feeble-minded and then provided free and good confinement facilities for the undesirable births. Both Reports proposed to stop this by segregating the idle, thriftless and mentally deficient who were already "making full use of their reproductive powers." A complementary change, in the Minority Report, would be to encourage breeding by people of good stock (and they had in mind working people rather than an elite) by providing financial support for raising children.

In many other ways the Minority report was quite radical requiring that action was needed to identify the causes of poverty and address them rather than accepting that pauperism was a normal outcome for working people and ameliorating it somewhat. It also emphasised the need for improvements in the environment – for the eugenic reason that is was no good conceiving good stock and then allowing it to degrade because of deprivation. But it is difficult not to conclude that progressive eugenic ideas were rather similar to everyone else's. It has been suggested that the Left was more cynical (or maybe more political) than this: the Fabians were simply adopting the vocabulary of eugenics (scientific, modern, middle-class) with no real commitment to its ideals.

## *THE EUGENICS SOCIETY*

By the time the Commission reported the Eugenics Education Society

# A BETTER SPECIES?

had come into existence. It was founded in 1907, on the initiative of Galton and the social reformer Sybil Gotto, with the specific purpose of pro-eugenic propagandising. Galton became its first President. The Society was to change its name to the Eugenics Society in 1923 and numbered among its members, at various times, John Maynard Keynes, Julian Huxley, R A Fisher and Marie Stopes, the family planning campaigner. Starting in London, the Society soon set up branches around the country.

In 1909 the Society started its journal *The Eugenics Review* – which was to be published for the next 60 years as a quarterly.

## POSITIVES AND NEGATIVES

There were two possible responses to the differential birth rate (assuming that a response was needed). The first was some variation on Galton's idea that increased the birth-rate among the threatened professional classes; the second was to find some way of decreasing the fertility of the lower classes: the unfit, unworthy and degenerate.

The first of these (which came to be called "positive eugenics") proved difficult: it depended essentially on persuasion rather than imposition. Economic obstacles to parenthood could be removed through grants and, perhaps, the tax system; the glory of motherhood could be promoted but the middle classes had discovered and were practising birth control (some 50% of them according to some estimates). Whatever, no visible progress had been made in increasing middle-class fertility by World War I.

Negative eugenics – decreasing the production of children by undesirables – began to seem a much more realistic prospect. Promoting birth control would be ineffective – these people were dissolute and irresponsible. The Commission on the Care and Control of the Feeble-minded when it reported in 1908 recommended segregation and the Poor Law reports were agreed that men and women should be separated. It was understood that the Government was sympathetic and would introduce legislation and Winston Churchill, the Home Secretary between February 1910 and October 1911, assured Parliament that a Bill was in preparation.[27] Churchill in fact took a personal interest in the Bill describing the "multiplication of the unfit" as a "very terrible danger to the race." Churchill changed places with Reginald McKenna[28] to become First Lord of the Admiralty in October 1911. The Government's will seemed to flag somewhat but, prompted by a Private Member's Bill and plagued by a well-orchestrated public campaign, the Mental Deficiency Bill was published and passed its second reading in 1912. This was not the end: objections from a determined group of backbenchers opposed to segregation and to restrictions on marriage (and to the eugenics arguments behind them) along with general concerns about its drafting and a shortage of Parliamentary time led to it being dropped.

A Mental Deficiency Bill was introduced by McKenna in 1913 but in this the eugenics arguments were hardly apparent and the provisions were all framed in terms of improvements for the sufferers themselves rather than (supposedly) for society and race. This passed into law and came into force on 1 April 1914 – although the eugenic purpose was far from explicit the eugenicists could claim a victory. It was, they said, the only social law where "the influence of heredity had been treated as a practical factor in determining its provisions." They clearly looked forward to more of the same.[29]

Another threat to society that might have responded to negative measures were habitual criminals. The Italian criminologist Cesare Lombardo claimed in 1876 (in his book *Uomo deliquente*) that his studies of such criminals (and their corpses) showed that they were throwbacks to an earlier and distinct (physically and mentally) race of man: *Homo deliquens*. They were innately criminal and so could not be reformed. However, given that they had distinct physical characteristics they could be identified. His uncompromising thesis was expanded in the following years and translated into several languages, including English in 1911. It led to insistent calls (from Galton among others) that persistent criminals should be separated and prevented from breeding on eugenic grounds; there was an expectation that criminals were particularly fertile with large families. The basis for Lombroso's theory – the supposed link between physical and criminal characters – was demolished by Charles Goring, a worker in Pearson's Eugenics Laboratory. So Lombardo's ideas were discredited – although they continued to have currency in the USA for some time – but what did emerge in their place was the notion that there was a connection between criminality and (what else!) feeble-mindedness. So the view that criminality was largely hereditary remained widespread – although statistics did show that criminals were in fact rather infertile. Repeated long spells behind bars reduced opportunity and many offenders found that their spouses had deserted them while they were inside. No specific eugenic measures were taken against criminals in the UK.

### *JUKES AND KALLIKAKS*

Dr Elisha Harris was an enquiring and innovative Registrar of Vital Statisitics for the New York City Board of Health – with a interest in a wide range of public health and welfare issues. When he applied his skills to the occurrence of family names in the prison population of county gaols he found some of them occurred more often than expected, particularly in

## A BETTER SPECIES?

Ulster County some 100 miles north of the city. Closer study of the Ulster County inmates allowed him to trace the origins of many of the inmates through six generations to a woman he named "Margaret, mother of criminals".

Illustration 8: An Ulster County Family (University at Albany, SUNY)

Ulster County, like others round about, had changed as the city grew and its transport tentacles spread into the region. Plutocrats and millionaires had built magnificent houses in the Catskill Mountains and the simply well-off had built comfortable week-end cottages. There were others who had settled and farmed or traded their way to respectability but Margaret came from the poor folk who had rejected the big city or failed in it and opted for a simple life of hunting and fishing. Perhaps idyllic when they settled there, the life became harder as the land and water on which they could freely hunt, fish and forage became smaller and smaller as the population grew. They were forced to turn to other ways of providing for themselves than the trap, gun and rod and while some no doubt found employment others became beggars and vagrants. Many – it seemed – became criminals.

Harris found 621 descendants of Margaret (and possibly her siblings); convicts, paupers, beggars, vagrants and criminals. Margaret, it was reckoned, had "cost the county hundreds of thousand of dollars"[30]. Harris and other commentators of the time assumed that his results demonstrated that criminality and other forms of degeneracy were inherited.

Harris was part of a small group that met to discuss social questions at the house of Richard Dugdale, a businessman and sometime sculptor, with an interest in the application of statistics to social issues. Harris invited Dugdale to follow up his work on records of county gaols in the region of Ulster and this he did in much more detail, interviewing prison staff and using the records of sheriffs' arrests and town poorhouses. He published his results in 1877 under the title *The Jukes*[31]: *A Study in Crime, Pauperism, Disease, and Hereditary*.

Dugdale's result broadly confirmed what Harris had found: there were families (or rather groups of families – we might call them clans) who seemed more liable to end up in pauperism or criminality than others and it appeared that the pattern was consistent with the traits being inherited. However Dugdale rejected the idea that this was direct genetic inheritance. In contrast to Harris and other interpreters of his results subsequently, he thought the major factor in keeping clans criminal and in poverty was their persisting, virtually inescapable, environment. It gave him quite a different perspective from that of the strict hereditarians: for him criminals could be reformed and the degeneracy caused by poor environments could be repaired. His liberal interpretation was lost to history and the story of the Jukes became synonymous with hopelessness and the stigmatisation of the extremely poor.

The Jukes study prompted several others of poor rural families and clans in the following decades: the Ishmaelites or Tribe of Ishmael(1888), the Smoky Pilgrims of Kansas (1897), the Nams of upstate New York (1912), the Hill Folk of Massachusetts (1912) and the Kallikaks of New Jersey[32]. The traits studied were various (criminality, pauperism and feeble-mindedness were clear favourites) but the message was almost always the same: the traits were ingrained in the genetic make-up and could not be changed. It was a bleak message for social reform.

### *ABA AND DAVENPORT*

The rediscovery of Mendel and the potential his ideas had for plant and animal breeding led to the formation of the American Breeders Association in 1903. The interests spread across the plant and animal kingdom with, naturally enough, an emphasis on improving working animals and edible plants and livestock. So there were committees on breeding of Carriage Horses, Cotton, Forage Crops, Horse Hybrids and Poultry for example. However, the interest in eugenics was such that, in

## A BETTER SPECIES?

1906, at the Second Annual Meeting, they set up a Committee on Eugenics under the chairmanship of David Starr Jordan.

Jordan was a naturalist (he made his name studying fish) who rose swiftly in the academic world to become, at age 34, President of Indiana University in 1885 and of the newly created Stanford University in 1905. Carlson suggests that there were two influences that led him towards eugenics. The first was discussions with Oscar McCulloch (who had "discovered" the Tribe of Ishmael) about them and Dugdale's Jukes where they shared a view that they were degenerate stocks and Jordan went along with McCulloch's likening of them to parasites in the animal kingdom. The second was drawn from his own experience in Europe. In 1881, 1883 and 1890, when he visited the village of Aosta in the Italian Alps, he found numerous mentally retarded and physically deformed people begging in the streets. These were sufferers from hypothyroidism – defects of the thyroid gland which reduce its ability to concentrate the element iodine with a terrible impact on their mental development and gave many of them characteristic goitres caused by the swollen glands.[33] Jordan saw the situation as a paradigm for misplaced charity and welfare: these people were being supported by the community, they married (sometimes one another) and had children, continuing the line of cretins (as they were sufferers were commonly known) without end. He was pleased to find when he went back in 1910 that the poor people had been segregated since his last visit and forbidden to marry and there was only one left. A success for "social castration".

He extended these views in predictable ways: pauperism was hereditary, the "venal, cowardly [and] ignorant" should be denied the vote, immigrants were generally the worst of their parent country. He was particularly interested in the eugenic effects of war. Rather than seeing war, as some did, as the way in which superior nations and races weeded out the weak ones and therefore as positive, he emphasised the negative. War, he claimed in several books between 1902 and 1915, eliminated the strong and the brave leaving the weaker survivors to father the next generation – he called it "the survival of the unfittest". He saw war as a major cause of the decline of races. He opposed war on other grounds than eugenic ones – expense for example – but it was probably the eugenic ones that determined the chilly reception he received when he addressed the Eugenic Education Society in London on 14 July 1913. Leonard Darwin, in the chair for the meeting, rather bluntly disagreed with Jordan and it cannot have helped that the venue was Sunderland House, the residence of the Duchess of Marlborough. The dukedom had been given to John Churchill, thought by many to have been Britain's greatest ever military commander, in 1702. Jordan's reception seems to have been even worse when he expressed his anti-war views to the German eugenics society; many of those present were members of the army general staff.

Jordan was, as well as being a noted eugenicist and peace campaigner, a respected university administrator – he left Stanford after 22 years in 1913 – and as an ichthyologist he had two genera and some 30 species of fish (and a research ship) named in his honour. He may yet, when Hollywood gets round to it, be more famous for his involvement in the murder of the wife of Leland Stanford, the founder of the University. Jane Stanford died an unpleasant death in the Moana Hotel, Waikiki, Hawaii on 28 February 1905. A local doctor present at her death pronounced it to be by strychnine poisoning (all the symptoms screamed strychnine) and this was quickly confirmed by an autopsy and an inquest – the poison was present in a potion she had drunk for indigestion. However, Jordan was on the scene within a few days and paid a local doctor with a dubious reputation $7000 to write a short report stating that the death was the result of a heart condition and it was this diagnosis that appeared in US continental newspapers and was generally accepted on the mainland. The real story lay buried for nearly 100 years until 2003 when a Stanford academic, W P Cutler, struck upon it and wrote a book – *The Mysterious Death of Jane Stanford*.

There were more suspicious circumstances. It emerged that the lady had avoided a similar attempt a few weeks before. Jordan's doctor had fled Hawaii very quickly after being paid. There had been friction between the deceased and Jordan because she interfered in the running of the university; one of Stanford's allies was dismissed shortly after her death. Perhaps Jordan's actions amounted to little more than a (successful) attempt to protect the University's reputation; Cutler left the verdict rather open. But that would surely be enough for Hollywood to indict Jordan.

The other members of the American Breeders Association committee were Alexander Graham Bell, Vernon Kellogg, Luther Burbank, Adolf Meyer, Herbert John Webber, Frederick Adams Woods, W E Castle and Charles B Davenport.

Bell was professionally a teacher of the deaf and a convinced oralist – he believed that the deaf should be taught to lip-read and speak rather than sign. It was experiments in visualising speech to facilitate this that led to his invention of the first practical telephone in 1876. He wrote extensively on deafness (and much else). His first eugenic publication was in 1883 when he presented *Upon the Formation of a Deaf Variety of the Human Race* at the National Academy of Sciences. In this he argued that since

## A BETTER SPECIES?

deafness was hereditary and deaf people were socialising in clubs and marrying, a deaf variety of humans would eventually arise. One of his reasons for favouring oralism was that it would counter this by making communication with the non-deaf easier.

One of his other enthusiasms was breeding sheep.

Eugenics became a life-long interest. He became Chairman of the Board of the Eugenics Record Office and wrote several papers between 1914 and 1919 on eugenic matters. In 1921 he was President of the Second International Eugenics Congress.

Kellogg was, like Jordan, a pacifist but changed his position when he spent time with some German staff officers in the early years of World War I. He wrote on the dyseugenic effects of war. Burbank, the eccentric plant breeder, was an enthusiastic eugenicist but was hampered by Lamarckian views: "heredity is only the sum of all past environment, in other words environment is the architect of heredity."[34] He believed that almost all hereditary conditions could be overcome by the appropriate environment so the weak and abnormal and even the mentally deficient should be improved. Ever, one feels, the plantsman, he placed great emphasis on good nutrition, fresh air and sunshine in the upbringing of children.

Adolf Meyer was perhaps America's leading psychiatrist of the first half of the 20th century. He was the first American ( he was actually born in Switzerland) to write on schizophrenia (or dementia praecox as it was then called) and was known for his empirical approach to psychiatry: he believed in collecting large amounts of data about all of his patients. Early support for eugenics (he was involved with the ERO) soured as he realised it had insubstantial scientific foundations. Herbert John Webber was a Professor of Plant Breeding at Cornell University and had rather more conventional views (for a eugenicist) that led him to regard negroes as inferior to whites, cross-breeding between races as highly undesirable and society as heading for catastrophe if degenerating tendencies were not kept in check. [35]

Frederick Adams Woods followed the thread that Galton had started – the study of eminent people to establish the heredity of genius – examining the cases of several thousand entries in the Hall of Fame. He supported the conclusions from this with a statistical analysis of royalty – a study population where environment was broadly uniform but genetic constitution varied widely. He managed to convince himself that genius and moral rectitude occurred in closely related groups – as did their opposites – and so were very likely to be hereditary.

# GENES, FLIES, BOMBS...

The Secretary of the committee was Charles Benedict Davenport, the Director of the Cold Spring Laboratory. The Lab had been founded in 1890 as a training college where teachers could come to study marine biology by the sea in the summer. Davenport, a Harvard biologist, had been part-time Director since 1898. But there was a dramatic expansion of the organisation when, at Davenport's suggestion, the Carnegie Institute set up a Station for Experimental Evolution there. Davenport, originally a biometrician under the influence of Galton and Pearson, became a convinced Mendelian and built it as a centre for experimental work with *Drosophila* and other creatures. However, after about 1907 his subject became Man and his personal interest in genetics was mainly directed through eugenics. In 1910 he was instrumental in setting up the Eugenics Record Office. His most important publication on eugenics may well have been *Heredity in Relation to Eugenics* published in 1911 but he retained an active interest in the subject beyond his retirement from the Laboratory in 1934. A ceaselessly active man, he published on a vast scale material some of which which was criticised later as hurried and poorly thought-out. In summarising his attitude to eugenics the author of his National Academy of Sciences obituary (he was elected in 1912), Oscar Riddle, made the following comment:

> His scientific background and associations gave him the prestige of an authority in the eyes of those inclined to accept his position on the social and political aspects of eugenics. But the opposition of many, instead of quickening the search for more accurate and convincing evidence, called forth a defensive attitude which led to exaggerated emphasis and dulled objective thinking.

His energy and enthusiasm finally killed him when, in his late 70s, he spent several weeks in January 1944 boiling the head of a killer whale that had beached off Long Island for its bones. He did this in a huge cauldron in a shed open to the bitter winter weather. Undeterred by the severe cold he developed, it took pneumonia and consequent death to stop him. He died on the 18 February 1944

## *EUGENICS RECORD OFFICE*

The Eugenics Record Office was founded by Davenport in 1910 with an endowment from Mary Harriman, the widow of E H Harriman, a railroad magnate. Davenport appointed Harry H Laughlin as its Superintendent. Alexander Graham Bell was the first chairman of its advisory board and T H Morgan was a founding member – although he later withdrew as his doubts about the nature of eugenics grew. The ERO's main area of study was the collection of human pedigrees. Dugdale's work on the Jukes was continued by Arthur Estabrook who also went on to look at the

# A BETTER SPECIES?

Ishmaelites and the mountain folk of the Appalachians. Others studied "aristogenic" pedigrees such as the eminent families of the Boston area.

Funding was provided by the Carnegie Institution from 1918 and in 1921 the ERO combined with the Laboratory to form the Carnegie Department of Genetics. Davenport became Director of this, with Laughlin as Assistant Director of the ERO. Davenport retired in 1934 and the ERO continued until 1939 when it was closed by Vannevar Bush. Laughlin had become heavily involved in the compulsory sterilization movement and in documenting the supposed inferior quality of Eastern European immigrants into the USA.

## *STERILIZATION USA*

Laughlin's name seems to have become synonymous with sterilization but it predated him as a medical and eugenic procedure by several decades. Oophorectomy was used as a treatment for hysteria – a condition thought to be connected with the ovaries – in the 1880s. A French Canadian woman was treated at age 33 to relieve her of melancholia and compulsive masturbation. Some thought it effective for nymphomania in mental patients.

Dr Walter Lindley, president of the Californian State Society of Medicine recommended in 1890 castration and spaying (a term meaning removal of the female sex organs now reserved for the procedure in animals) for rapists, wife-beaters, idiots, murderers and some of the insane. He thought it should be compulsory. It is not really clear whether he had real eugenic motives in mind but the inclusion of idiots suggests he did. Some considered castration as a possible punishment for rape: the general and apparent feminisation that followed would make the guilty outcasts. Any who reformed as a result would be, he thought, welcomed into church choirs. Others pointed out that compulsory "unsexing" of criminals would be more humane than long terms of imprisonment and prevent them reproducing their kind.

The first large-scale programme of oophorectomy was planned for an insane asylum in Pennsylvania but was stopped after the fifth patient died during surgery. The plans of the Texan physician F E Daniels for forced castration of sexual criminals (although he would obviously have liked a much wider congregation that included the syphilitic, drunks and the insane) were thwarted. He included rape, sodomy, bestiality and habitual masturbation as reasons for castration but no law was enacted. Carlson points out that one of the reasons for opposition to compulsory sterilization was the risks it carried and (particularly for male castration) the mutilation it involved. A procedure was to come along at the end of the century that was less traumatic, more acceptable.

The vasectomy had been used in England in 1893 as a procedure that might stop the swelling of the prostate gland and by 1899 it was being used in the USA by Albert J Ochsner, an eminent and influential surgeon. He advocated vasectomy as an alternative to castration for criminals, perverts and degenerates to stop them breeding their own kind but seems not have have done any for these reasons. It was Dr Harry Sharp who first did this.

Sharp was physician in the Jefferson reformatory in Indiana from 1895 to 1910 and in 1899 he started vasectomies as a treatment for patients who compulsively masturbated. Over the next eight years he performed the procedure on 176 men. It was always, he claimed, voluntary but this was against a background of him issuing dire warnings to the men about the physical and mental consequences if they carried on abusing themselves. The programme was such a success in Sharp's eyes that he promoted the idea of compulsory sterilization of degenerates with the state Governor with some persistence and he was rewarded with the first State law allowing just this in 1907.

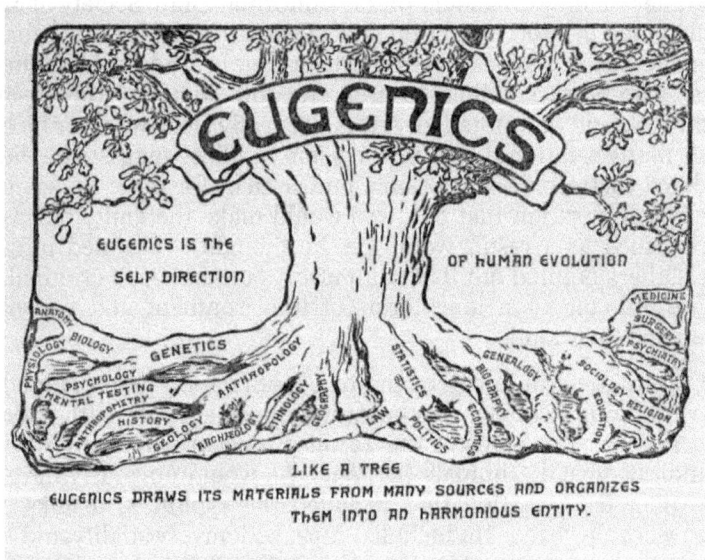

Illustration 9: The Eugenics Tree 1921

Indiana was not the first state to consider a law. Michigan had debated one in 1897; one for Pennsylvania failed in 1905 only because the

# A BETTER SPECIES?

Governor declined to sign it on the grounds that it was unconstitutional. In fact the Indiana law had a rather short effective life as a new Governor there forbade Sharp from undertaking vasectomies from 1910. The law stayed, unused, until it was removed in 1921 as breaching the 14$^{th}$ Amendment. Revised and re-instated in 1927 (and expanded in 1931) the law was used to sterilise rather more than 2000 inmates between then and the early 1960s. Seldom used after that, the law was repealed in 1974. However other states did introduce sterilization laws within a few years of Indiana. California, Connecticut and Washington passed them in 1909; Nevada, New Jersey and Iowa two years later and New York in 1912. Some were aimed at the mentally defective and insane, some at criminals, particularly those convicted of sex crimes – some were punitive and others had eugenic motives. Not all of them survived for more than a few years. It was this that really brought Harry Laughlin into the picture. Laughlin was rather too enthusiastic a promoter of sterilization for both Davenport (who worried that separation of act from consequences might cause even greater profligacy among the degenerates) and the Carnegie Institute (which valued research and deplored promotion). Laughlin therefore, while remaining Superintendent (and later Assistant Director) of the ERO, increasingly took on the role of consultant to those promoting sterilization. In 1922 he published a model law for compulsory sterilization which, he thought, avoided many of the flaws that had led to earlier laws being rejected or rescinded.

## *THE PEITHOLOGIAN LECTURE*

So it was that on 24 March 1910, a sheaf of notes in his hand, this short, slight young man walked past the imposing Low Memorial Library with its massive dome (still the largest all-granite dome in the USA) and into West Hall heading towards Room 404 to deliver an address on eugenics. He would no doubt have been aware of a small irony. The Columbia Morningside Heights campus incorporated the former Bloomingdale Insane Asylum. Most of the Asylum buildings had been demolished to make way for the University but three remained. West Hall was one of them.[36]

West Hall may have since gone but Muller's draft lecture notes survive in the Lilly Library in Indiana, so we have a very good idea of what he said. They also suggest that Muller was revising them right up to the last minute. The first 44 pages are written neatly out in a notebook with hardly a correction, the next nine are pages torn from a notebook with many amendments. The final three are ragged scraps of paper in a hurried, looser, hand. In total there are around 10,000 words.

In the first part of the talk he sets out the then-current views of heredity

# GENES, FLIES, BOMBS...

and its mechanisms and of evolution. In the second part he considers the implications for mankind and – moving on to his eugenic theme – argues that civilisation reduces the effects of natural selection. The weak are no longer removed and degeneration is the inevitable result. Indeed undesirables actually flourish.

> Improvident, ignorant, reckless and selfish people, the dregs of the cities, are breeding like veritable disease-germs, and the people of the cultured classes are not maintaining their numbers. p31

Races have slowly degenerated and often been swept away by barbarians still honed by natural selection. The Greeks and Romans declined and fell as did (and this now looks decidedly odd) the Chinese and Arabs.

Education is not the answer. No-one can be educated beyond a certain limit: "a dog can never learn to figure", as he puts it. The lower forms must be eliminated; the multiplication of the higher form accelerated(p28). The unfit must be eliminated as they would be by natural selection, but humanely – by keeping them from being born.

He quotes with approval the Indiana law providing for the vasectomy of confirmed criminals and the feeble-minded. It marked the beginning of an attack on the "momentous problem". This approach should not be extended to the poor, removing their ability to breed, because the economic system is imperfect and wealth is not a criterion of worthiness. Under a more ideal system, such as socialism, wealth might provide a much easier method of valuation of individuals. For the moment it might be possible to select those to be sterilised on the basis of their intelligence and moral character – provided scientific methods were used.(p37)

The accelerated multiplication of the higher forms is not considered in detail but the young Muller becomes quite rapturous about the possibilities of the "Eugenic System" (a term he introduces towards the end of the address): progress would be unlimited, "world-old" problems would be solved, anything would be conquerable.

The notes now enter their heavily-corrected sections and discuss consciousness, what he calls the satisfiers of consciousness, happiness as the aim of consciousness and Love. Eugenics poses no problems for those who believe in Platonic Love and should not do so for those interested in the "other" sort since gratification would be possible without offspring.

In a prescient prediction of his own future life's work:

> It may not be necessary to do as much breeding as I seem to have indicated in order to [achieve] the desired result. We shall probably find a way of inducing variations, so that the breeding process may

be much abridged. [It] may even be found possible to <u>control</u> variations and thus consciously and [purposefully] to <u>produce whatever</u> character we wish [without resorting] to breeding at all.[37]

All this must be informed by science: biology, heredity, physiology, psychology must all play their part. Then man will reach his "third age" in which the universe will "shake at his glance and forming itself into his image countless times repeated, shall shape itself into a living arch of victory for his march."

His final note after this exultation is written in a smaller, more controlled, hand:

> I fear that I've exhausted you tho not the subject. Please allow me to extend my apologies.

The final page of the file in the Lilly is a sketch showing creatures, mainly men or monkeys but with a few dog-like animals, struggling upwards towards a vent from what looks like a cavern. Some creatures, let's call them humanoids, crawl away from the vent in the land above (or maybe fly), having escaped, others seem to be waving goodbye and in free-fall in the darkness below. A Blake-like image inhabited by characters from James Thurber cartoons, it is titled

EUGENICS: THE JOURNEY TO THE PROMISED LANDS NO MAN HIMSELF MAY REACH.

It could well have been called "Out of the Night". But that's another, later, story.

## THE USA AFTER 1910

The Virginia Sterilization Act of 1924 seemed well-framed. Carefully restricted to sufferers from "hereditary forms of insanity that are recurrent, idiocy, imbecility, feeble-mindedness or epilepsy" and without the punitive elements that had led to the striking down of other laws, it had safeguards intended to protect the constitutional rights of the individual.[38] It seemed to meet the needs of due process. Laughlin was impressed and suggested that there should be legal test case in the courts to establish the law's constitutionality. He suggested to the Superintendent of the State Colony for Epileptics and Feeble-minded in Madison Heights, near Lynchburg VA, a Dr Albert S Priddy[39], that he should pick a patient and prepare an Order for their sterilization under the Act. Priddy selected Miss Carrie Buck and the Order to sterilize her was appealed in the Amhurst County Circuit Court by Buck's guardian, appointed by the Court, Mr Robert G Shelton. If this court declared the law unconstitutional, Priddy would appeal to the state Court of Appeals; if the law was upheld it was agreed that Shelton would appeal. Again, whatever the outcome at state level there would be a further appeal to the US Supreme Court. It was a carefully prepared test of the law.

In his petition to his Board of Governors Priddy summed up the case for sterilization:

> The said Carrie Buck, by reason of the laws of heredity, would in all probability, if permitted to bear children, transmit to her offspring some form of mental defectiveness by which the offspring would, in view of professional experience and teaching, develop some form of mental disease or defect such as feeble-mindedness, insanity or epilepsy, and by reason of her anti-social conduct and mental defectiveness, she is unfit to exercise the proper duties of motherhood. The said Carrie Buck is possessed of good physical health and strength and if by sterilization she be made incapable of child-bearing could leave the Colony and enjoy the liberty and blessings of outdoor life, become self-supporting and thereby relieve the Commonwealth of Virginia of the burden of the support of her under custodial care, in a State institution for mental defectives during the period of child-bearing and the said Carrie Buck desires that the said operation be performed.[40]

Carrie's case was surely a sad one: a poor white, she was born illegitimately to a mother who was quite possibly mentally retarded and who had maybe taken at some point to prostitution, had syphilis and produced three other children by different fathers. The mother, Emma, had been abandoned by her husband and was committed to the Virginia Colony as feeble-minded – with a mental age of less than eight – and an immoral person. Carrie was placed with foster parents and she helped around the house but, after she became pregnant when she was 17 – probably as a result of being raped – she was herself committed to the Colony as feeble-minded (with a mental age on nine) and promiscuous.

Harry Laughlin summed up the family in his testimony to the Circuit court:

> These people belong to the shiftless, ignorant and worthless class of anti-social whites of the South.[41]

The results of the three appeals at county, state and federal level were essentially the same. The Virginia law was constitutional and it and well-formed and the process it required had been followed. It became known as *Buck v Bell* because Superintendent Priddy died soon after the State appeal and the case was taken up by his successor John H Bell. The judgement of the US Supreme Court in 1927 (by a majority of 8-1) legitimised the possibility of legal compulsory sterilization of qualifying inmates across the USA. It just required well-formed laws.

The judgement was written by Supreme Court judge Oliver Wendell Holmes Jr and includes the following passage:

> We have seen more than once that the public welfare may call upon the best citizens for their lives. It would be strange if it could not call upon those who already sap the strength of the State for these lesser sacrifices, often not felt to be such by those concerned, in order to prevent our being swamped with incompetence. It is better for all the world, if instead of waiting to execute degenerate offspring for crime, or to let them starve for their imbecility, society can prevent those who are manifestly unfit from continuing their kind. The principle that sustains compulsory vaccination is broad enough to cover cutting the Fallopian tubes. Three generations of imbeciles are enough

Carrie Buck was sterilized by salpingectomy in October 1927 not long after the Supreme Court ruling – her Fallopian tubes were tied off. She left the Colony shortly after and worked as a domestic servant. She married twice and died in a nursing home in 1983. Her daughter Vivian, pronounced feeble-minded at the various hearings, did well at school but died in 1932, aged eight, from the complications of an intestinal disease.

The *Buck v Bell* ruling resulted in new and revised laws so that by 1942 some 30 states had them on their statute book. With the fresh confidence that the ruling gave to eugenicists, the 12,000 operations that had been performed on the mentally ill and deficient to 1931 increased to over 38,000 by 1941. Over 15,000 of these operations had been performed in California. By the 1960s the rates of sterilization declined quite dramatically. By then a total of over 60,000 patients has been operated on in the US with more than 20,000 in California, 7,000 in Virginia and 6,000 in North Carolina.

By 1942 some 13 states had laws permitting the sterilization of habitual criminals. There seems little evidence of how many such operations were performed but they are generally thought to be a small fraction of those done on the mentally ill or deficient. However, it is known that over 600 operations were performed in San Quentin alone by then.[42] The year is significant because it was then, in *Skinner v Oklahoma*, that the US Supreme Court struck down the 1935 Oklahoma Habitual Criminal Sterilization Act. It found that Skinner, who stole chickens, was liable to be sterilized while those guilty of white-collar crimes were not – a violation of the 14th Amendment's Equal Protection. *Skinner v Oklahoma* seems to have put some kind of brake on the enthusiasm for sterilization engendered by *Buck v Bell*.

## *AMERICAN EUGENICS SOCIETY*

The American Eugenics Society was founded in 1922, after the International Congress in New York in 1921. Originally the Eugenics Committee of the USA then the Eugenics Society of America it became the AES in 1925.

Although eugenics had a long history, until the AES came along there was no national focus for it. For a decade or so it flourished as an organisation promoting eugenic ideas through public lectures, conferences and, notably, exhibitions at state and county fairs. A feature of many of the latter was the "Fitter Families for Future Firesides Contest".

One common exhibition display featured lights that flashed to signal a birth. Every 15 seconds a person is born in the US. Every 48 seconds a person is born who will never have a mental age of more than eight. Every 50 seconds someone is sent to jail – and very few normal people go to jail. Every 7½ minutes a person capable of creative work and be fit for leadership is born.

A light flashed every 15 seconds indicating each $100 caring for people with bad heredity: the insane, feeble-minded, criminals and other defectives.

The Fitter Families Contest itself started with contestant families submitting a short record of family traits and this was supplemented by physical and psychological checks performed by a team of doctors on each member of the family. Each family member was given a grade of "eugenic health" ranging from A+ downwards; anyone with B+ or better received a bronze medal[43] inscribed with the legend "Yea, I have a goodly heritage." The families were divided into three categories (small with one child, medium with two to four and large, with five or more) and then, in each category, the family with the best average score across its members was given a silver trophy. The winners were invariably white with a northern European pedigree.

Another vehicle for the eugenic message was the sermon-writing contest. From 1926, for a few years, the AES organised contests for the best sermons on a eugenic theme with prizes of several hundred dollars. For the 1926 contest 180 adverts were placed in religious and other publications inviting ministers, priests, rabbis and theology students to preach a sermon on the topic "Religion and eugenics: Does the church

# A BETTER SPECIES?

have any responsibility for improving the human stock?" As many as 200 sermons were preached and 70 of them were submitted to the AES for judgement by Davenport, Prof William Lionel Phelps of Yale and Ozora Stearns Davis, the President of the Chicago Theological Seminary. The winner was the Rev Phillips Endecott Osgood, the rector of St Mark's Church in Minneapolis. His sermon "The Refiner's Fire" combined "a Social Gospel vision of ushering in the Kingdom with practical recommendations for eugenic legislation."[44]

The good reverend turned out to be a leading member of the Minnesota Eugenics Society and a fervent lobbyist for a tightening of Minnesota's sterilization laws to make procedure compulsory for the feeble-minded in state institutions.

## HUMAN BETTERMENT FOUNDATION

The Human Betterment Foundation was created in 1928 by Ezra Seymour Gosney, who had made a fortune in citrus fruit growing. Its origins lie a few years earlier when, prompted by reading Laughlin's book on sterilization and a meeting with him, Gosney hired biologist Paul Popenoe, a former editor of the American Breeders' Magazine (and Journal of Heredity as it became), to compile data about the California sterilization programme. In 1929 this was published as a book by Gosney and Popenoe as *"Sterilization for Human Betterment: a summary of the results of 6000 operations in California 1909-1929"*. The book and many other publications by the HBF promoted eugenics in general and sterilization in particular for the next decade. They were influential not only in the USA but internationally. The Nazis in Germany seem to have learned a great deal from the American experience and Popenoe could write in 1934 of their progress in sterilization:

> But the Nazis seem, as this scientific leadership becomes more and more prominent in their councils, to be avoiding the misplaced emphasis of their earlier pronouncements on questions of race, and to be proceeding toward a policy that will accord with the best thought of eugenicists in all civilized countries. In any case, the present German government has given the first example in modern times of an administration based frankly and determinedly on the principles of eugenics. It has thus posed the question in a way that no other people can ignore.[45]

With support from Gosney, Popenoe founded the American Institute of Family Relations – a marriage counselling service in Los Angeles – in 1930 on the basis that sound marriages made for fitter offspring. As the public interest in eugenics declined after the war, he segued across to a

# GENES, FLIES, BOMBS...

very successful new career in marriage guidance.

## DEVELOPMENTS IN THE UK

A Private Member's Bill[46] was introduced by A G Church (a Labour MP) to legalise voluntary sterilization but defeated.

> We realise that we have to convert a large section of the people of this country to a full appreciation of what we propose to do with those who are in every way a burden to their parents, a misery to themselves and in my opinion a menace to the social life of the community.

Although intended to apply to cases where patients themselves consented to sterilization, the Bill was explicitly a step towards compulsory sterilization.

> This Bill is merely a first step in order that the community as a whole should be able to make an experiment on a small scale so that later on we may have the benefit of the results and experience gained in order to come to conclusions before bringing in a Bill for the compulsory sterilisation of the unfit.

There was reassurance that the operation was simple and safe:

> The operation is not the mutilation usually associated with sterilisation. I need not go into the details, except to say that it is devised in order that the emotional life of the patient shall not be affected in the slightest degree, but that the patient shall be effectively prevented from bringing children into the world. There are various operations on the male. There is a very simple operation which is done in five minutes, usually with a local anaesthetic. There is no danger attached to it. Out of 5,000 odd cases dealt with in this way in California, there has been only one death, and that was under the anaesthetic.

The response of Dr Hyacinth Morgan (1885-1956) MP for Camberwell North at the time, a London GP and founder of the Socialist Medical Association was unequivocal:

> I rise to ask the House not to give leave for the introduction of this Bill. The House has heard a harrowing tale which is mostly moonshine. The Bill is said to be in advance of public opinion, but it is really in advance of common sense and ordinary sanity. With regard to mental defectives there is said to be an increase rising crescendo in geometrical progression to overwhelm the world in an avalanche of mental backwardness, and to lure the progressive

# A BETTER SPECIES?

world headlong into an abyss of degenerate civilisation.

> If once the principle of maiming or mutilation is admitted, not for the benefit or health of the individual but for the good of others or the State acting for others, there is no brake to sliding down the slippery slope leading to the swamp of State penalisation, where we may get rid of all those obnoxious to the State. Those preaching subversive doctrines may have their tongues cut out. Those writing subversive doctrines may have their hands cut off. The State (those temporarily in power) are the dictators of limb and life. The eugenist upon a pinnacle of intellectual snobbery, looking down upon the less fortunate mental defective, may gradually raise the standard of mental deficiency and push more and more citizens into the maelstrom of the mentally-maimed.

Dr Morgan continued:

> I submit that this is class legislation. In Europe there are Monarchies and dynasties riddled with haemophilia, a disease transmitted by the females but affecting the males. It causes bleeding on the slightest provocation or injury. I have never yet heard one expert speak of the advantage of sterilisation in the case of these royalties, and I submit that for these poor, helpless, mentally-maimed people the ordinary average mental defective person, we should adopt the other methods of segregation and socialisation. If we take the long view and not the short view, ultimately mental deficiency, instead of being a terror and a menace, may yield good results to proper methods. I ask this House to refuse to give leave to introduce this pagan, anti-democratic, anti-Christian, unethical Bill.

The Bill was rejected.

## DECLINE & CHANGE OF EUGENICS

The popular enthusiasm for eugenics in the USA ran high through the 20s. Scott Fitzgerald wrote and performed a ditty, *Love or Eugenics,* for the 1914 Princeton show and referred to it in his iconic 1925 novel *The Great Gatsby* – making the rather brutish character Tom Buchanan a supporter of eugenics and a racist[47]. It was supported by prominent politicians such as Winston Churchill and Theodore Roosevelt, by radicals like the Webbs and H G Wells and George Bernard Shaw in Britain

In the early years of the eugenics movement many professional biologists and geneticists were enthusiastic supporters of the eugenics movement and no small number of the prominent among them supported it tacitly. The new religion made their science popular and seemingly important. As Kevles says "They warmed easily to their priestly role."[48] However there were always concerns from academics. T H Morgan resigned from the American Breeders' Association condemning (privately) the quality of its science. Others drifted away disturbed by the racist aspects or, increasingly, suspicious of the scientific foundations of the creed.

The casual identification of traits as simply heritable that characterised some of the research was rejected: there was an increasing recognition that many traits were polygenic in nature and, at the time, near unfathomable. There was also a debate to be had about the effectiveness of sterilization as a way of eliminating undesirables; population genetics showed that recessive genes could persist for a very long time if only the homozygotes who displayed the trait were prevented from reproducing. If the traits were polygenic it just made things worse.

It also seemed by the 1920s that there was little evidence for the deterioration of the populace that eugenics had predicted. In the USA, the IQ tests that had been interpreted to mean decline were attacked on the basis that social and environmental factors were at least as important as genetic ones. In the UK, large-scale tests of Scottish children showed a slight rise in scores between 1932 and 1947. If there had been an increase in the number of people classified as mentally deficient, it was more because of a change in classification and modes of care than any change in the intellectual structure of society. The notion of race and its link to innate ability faded – negroes and immigrants were seen less as undesirable inferiors.

Generally environmental factors, such as environment, nutrition, poverty, class and education were increasingly recognised as important. The simpler messages of eugenics (good begets good; bad begets bad) were less acceptable. Where the first two international congresses attracted large numbers (more than 400 for the first one), by the third (and last) in New York in 1932 barely 100 people turned up. In the USA the ERO continued until 1939 when the Rockefeller cut off its funding. By now eugenic sterilization, even if still legal, was hardly enforced.

The papal encyclical *Casti connubii* was issued in December 1930 by Pope Pius XI. It supported the sanctity of marriage and opposed birth control and eugenic sterilization largely as a response to the Anglican Lambeth Conference's support that year for birth control. It seems to some not to have opposed the principle of eugenics but rather the negative methods of its implementation. Quite how influential it was among non-

# A BETTER SPECIES?

Catholics is a matter, still, of debate.

The mainline, scientifically-flawed eugenics of Davenport, Laughlin, Major Darwin and Pearson more or less collapsed before the Second World War. But it was not quite the end of eugenics: a more science-friendly and (perhaps) more compassionate version had emerged. Kevles calls it "reform eugenics". It was here that Muller took his place.

## MULLER AND EUGENICS

After the Peithologian lecture Muller was almost silent on eugenics for over 20 years. There was a public talk at the Rice Institute in 1919 and another at the University of Chicago in 1925 but nothing really significant until 1932 when he spoke, to Davenport's dismay, at the International Congress in 1932.

## DOMINANCE PAPER

Muller's paper *Dominance of Economics over Eugenics* at the Eugenics Congress 23 August 1932 was expected to last 60 minutes but his slot was cut to 10 minutes by the Congress organisers when they saw the paper. An extract from the abstract of the published version of the paper[49] gives some idea why.

> It has been shown that eugenics under capitalism involves several serious contradictions:
>
> .... individual economic considerations rather than considerations of the genetic worth of the future generations must in the main govern human reproduction, in so far as the latter is voluntary at all, and eugenics must remain an idle dream.
>
> ...to justify the existence of the gross economic and social inequalities between classes, races and individuals, arising under our present economic system, it has been necessary for the apologists of this system to put forward the naive doctrine that the economically dominant classes, races and individuals are genetically superior. Such scientific evidence as is available fails to support this contention,
>
> Our economic system....inculcates predatory rather than constructive ideals. In consequence, the ideal set of characteristics which most present-day eugenists and the population at large would set up as a eugenic goal, is far from the type which would be considered most desirable in a well-ordered society.
>
> It is advantageous for the dominant classes, under our present system, to foster a set of archaic superstitions and taboos, and these

> are directly antagonistic to rational, civilized practises regarding sex and reproduction.
>
> ...the impeding radical changes in our economic order are prerequisite to a genuine, functioning eugenics.

While few may have agreed with the socialist programme that Muller had in mind, this was a damaging attack on the eugenics idea, emphasising as it did the importance of environment over eugenic issues. It may not have been the mortal blow to eugenics in the US that some commentators have claimed[50] (the Depression made people only too aware of the profound effects of economics) but it signalled its end.

### OUT OF THE NIGHT

Muller published *Out of the Night*[51] in 1935. In many ways a simple extension of the thoughts he set out in his Peithologian address so many years before, these now had an explicitly sociological theme. The thinking started too, much more clearly, with a technical genetics rationale – based on the notion of genetic load:

> The crux of the matter is that new mutations, though rare, are continually occurring, and cannot be prevented, and that, in any population whatever, the mutations which give rise to "defective" or "pathological" traits are relatively far more abundant than the "beneficial" mutations. We cannot, therefore, by any possibility escape the conclusion that the process of mutation will in time cause a gradual heaping up of undesirable traits of all sorts in any group of animals or plants, if individuals having the defective genes are allowed to multiply merely at the some rate as the others. One after another the remaining "normal" genes will themselves "go bad" through mutation, and then survive in their changed form. So long as this condition holds, then a biological disorganisation will take place, and will necessarily continue without limit, or until the species disappears.

The answer was "some kind of thoroughgoing eugenics" (he dismisses the contemporary eugenics as perverted) and that required "urgently a development of sympathy, a social idealism, a degree of co-operation higher than that which now exists"...."a higher form"(p54).

The sterilization of the unfit and feeble-minded would not be effective (p98) so the focus seemed to be on how to get the "best" people to have more children. For women this meant removing the burdens of pregnancy and parenthood through birth control, abortion, making childbirth less painful, reducing chronic infant illness, making child-rearing easier and

## A BETTER SPECIES?

eliminating discrimination against mothers working. There should also be investment in the development of means to "extend the reproductive potencies of females possessing characters particularly excellent, without thereby interfering with their personal lives" (p137). He particularly means ways of transplanting eggs from these favoured women so they could develop and be brought to term outside the mothers' bodies – surrogacy. It was, he recognised, something for the rather distant future.

The breeding from superior men would be much easier and could be done even then, in 1935. Men produced billions of sperm and artificial insemination was quite possible producing, if required, almost boundless offspring of desirables. Perhaps the only problem was who the breeding stock should be. Muller suggested Newton, Leonardo, Pasteur, Beethoven, Omar Khayyam, Pushkin, Sun Yat Sen and Marx[52] as good examples. He was concerned that, with the current nature of society, people would prefer "a maximum number of Billy Sundays, Valentinos, John L. Sullivans, Huey Longs, even Al Capones"[53](p142).

The new society based on altruism, cooperation and responsibility – eschewing the profit motive and the selfishness that goes with it – would be needed before this ambitious programme could succeed and then we could aim for the third phase of life (harking back to the Peithologian ideas):

> And in the long third phase it will reach down into the secret places of the great universe of its own nature and, by the aid of its ever-growing intelligence and co-operation, shape itself into an increasingly sublime creation – a being beside which the mythical divinities of the past will seem more and more ridiculous, and which, setting its own marvellous inner powers against the brute Goliath of the suns and planets, challenges them to contest.(p157)

*Out of the Night* sold around one thousand copies in the USA in 1935. When it was published in the following year in Britain it sold over ten times as many. The UK publication was arranged by Herbert Brewer, an English amateur eugenicist and socialist.

### *EUTELEGENESIS*

Brewer had coined the term eutelegenesis as a title for a paper in Eugenics Review[54]. It referred to a practice that Brewer thought might promptly and quickly (in just a few generations) advance eugenics: artificial insemination of consenting women with the sperm of the best (and the greatest) men. It was an idea that Muller approved of and he mentioned Brewer and the idea in *Out of the Night*.

# GENES, FLIES, BOMBS...

Artificial insemination of animals had been practised for some time and the idea of applying it to humans was not repulsive. Brewer could report interest in the procedure already among childless couples in the USA. It clinically separated procreation from emotion and love (and was hence a eugenicists dream) and was hardly even unpleasant. All it seemed to need was some body to organise and regulate the process and an effective eugenics programme could get started. It would, of course, be voluntary but Brewer saw no shortage of suitable candidate sperm and wombs.

## THE MANIFESTO

In 1939 F A E Crew, Muller, Huxley, Haldane, Lancelot Hogben and others signed what became called "The Geneticists' Manifesto" at the Seventh International Congress of Genetics at Edinburgh[55]. This endorsed genetic improvement but by voluntary and informed means (rather than for example compulsory sterilization) in a society organised for production "primarily for the benefit of consumers and workers" and dedicated to the needs of mothers and children. Their protests against the eugenics movement condemned the extremes of negative eugenics and rejected racist attitudes inherent in much eugenics but were more a protest against the nature of society than against the eugenic principle itself. Indeed Muller in 1932 and the signers of the Manifesto were complaining that contemporary society was essentially incapable of achieving genetic improvement.

Various eugenic programmes were in place around the world. Sweden passed several laws, the first in 1927, for both voluntary and compulsory sterilization. Racial purity was one reason but feeblemindedness and genetic disorders were others. Estimates put the total number of young women operated on as around 60,000 with maybe as many as 20,000 being forcibly sterilised. The other Nordic countries followed the same pattern but the number of sterilizations was far lower.

Alberta in Canada introduced a compulsory sterilization law in 1928 and as many as 2000 may have been operated on. Laws were passed in Switzerland and in Japan there were campaigns of both positive and negative eugenics. Many other countries passed laws that may be construed to have had eugenic intent. However it was in Nazi Germany that the eugenic doctrine had its most complete and brutal application.

## GERMANY

Eugenics in Germany grew as a doctrine more or less in line with how it developed in other countries. There were adherents (some prominent in the movement) of the Aryan or Nordic creed but most eugenicists were no more racist than anywhere else in the world. Their primary concerns were with the degenerate and unfit who were thought both a burden and a threat to Germany. Even Fritz Lenz, the most prominent (and later reviled) German eugenicist of the 1920s and early 30s, seemed more concerned with ensuring the continuation of the educated middle-classes than anything else. A supporter of the Nordic-Aryan myth and an anti-Semite surely but then so were many others of his class – and not just in Germany.[56]

The coming of the Nazis in 1933 and the prompt passing of racial hygiene laws was welcomed by eugenicists in the country – and elsewhere round the world. The 1933 Law for the Prevention of Genetically Diseased Offspring, 1933 (The Sterilization Law)

> Anyone with hereditary disease may be rendered sterile by surgical means, when, according to medical experience, it is highly probably that the offspring of such person will suffer from severe inherited mental or bodily disorders.

Some of the diseases that might lead to sterilization were listed:

> 1. Congenital feeble-mindedness. 2. Schizophrenia (dementia praecox). 3. Manic-depressive insanity. 4. Inherited epilepsy. 5. Huntington's chorea. 6. Hereditary blindness. 7. Hereditary deafness. 8. Severe hereditary malformation.
>
> Also, anyone with severe alcoholism may be sterilized

Published official estimates put the number of people in Germany who suffered from these conditions at nearly 400,000.[57] Some interpreted this to mean that this number would be sterilized. Others pointed out that this ignored the administrative structure, based on independent Eugenic Courts, that would fairly implement and regulate the law; the number surely would be lower.

The move was widely applauded by eugenicists around the world – even by those who had been disturbed by some of the racist language that had been used by the regime. They were relieved that Hitler had embraced some prominent eugenicists such as Lenz. As Paul Popenoe wrote at the time:

> At best, mistakes will be inevitable. But the Nazis seem, as this scientific leadership becomes more and more prominent in their councils, to be avoiding the misplaced emphasis of their earlier pronouncements on questions of race, and to be proceeding toward a policy that will accord with the best thought of eugenists in all civilized countries.

## GENES, FLIES, BOMBS...

> In any case, the present German government has given the first example in modern times of an administration based frankly and determinedly on the principles of eugenics. It has thus posed the question in a way that no other people can ignore.[58]

The support for Germany endured with some up to 1939 and beyond but significant figures and organisations (such as the UK Eugenics Society under the influence of Huxley and Blacker) had grave doubts about the Nazi programme. The racist nature of this became much more apparent with the passing of the two so-called Nuremberg Laws in 1935. The Law for the Protection of German Blood and German Honour forbade marriages between Jews and non-Jewish Germans and the Reich Citizenship Law removed citizenship rights from those deemed to be not of German blood.

By the time Nazi eugenics extended to mass killing, even if the existence of the death camps was not known, the brutality of the regime was apparent and all but the very few outside Germany rejected it in horror.

After the war much less was heard of negative eugenics. There had been growing opposition to sterilization programmes for eugenic reasons, for both moral and scientific reasons, but the Nazis were a warning of where negative eugenics might lead. It emerged that Hitler had been encouraged by the American sterilization programme – particularly by the Californian experience. The sterilization programmes around the world faltered and, if laws remained on the statute book, they fell into disuse. The anti-miscegenation marriage laws passed earlier by many US states were repealed after the war. That in Virginia was, in 1967, the last to go after a Supreme Court decision.

### GERMINAL CHOICE

Muller's interest in eugenics remained after the war. In one of his last publications, prepared for the Ciba Foundation Symposium in London in London in 1962[59], it was still motivated by a familiar belief that "genetically based ability and reproduction rates are today negatively correlated." In this he reaffirmed his belief in eutelegenesis as a practical and already-available solution. Banks of sperm cells with contributions from the "most outstanding in regard to valuable characteristics of mind, heart and body" would be kept frozen and then, after a delay of perhaps 20 years, made available to women who wanted (and maybe deserved) them to conceive children.

> When the choices are not imposed but voluntary and democratic, the sound values common to humanity nearly everywhere are bound to exert the predominant influence in guiding the directions

# A BETTER SPECIES?

> of choice. Practically all peoples venerate creativity, wisdom, brotherliness, loving-kindness, perceptivity, expressivity, joy of life, fortitude, vigour, longevity. If presented with the opportunity to have their children approach nearer to such goals than they could do themselves, they will not turn down this golden chance, and the next generation, thus benefited, will be able to choose better than they did. (p260)

The Symposium was opened by Julian Huxley who, in a wide-ranging talk, re-emphasised his profound belief in eugenics:

> The improvement of human genetic quality by eugenic methods would take a great load of suffering and frustration of off the shoulders of evolving humanity, and would much increase both enjoyment and efficiency. Let me give one example. The general level of genetic intelligence could theoretically be raised by eugenic selection; and even a slight rise in its average level would give a marked increase in the number of the outstandingly intelligent and capable people needed to run our increasingly complex societies. Thus a 1-5 per cent increase in mean genetic intelligence quotient (I.Q.) from 100 to 101-5, would increase the production of those with an I.Q. of 160 and over by about 50 per cent.
>
> How to implement a eugenic policy in practice is another matter. The effects of merely encouraging potentially well-endowed individuals to have more children, and vice versa, would be much too slow for modern psychosocial evolution.
>
> Eugenics will eventually have to have recourse to methods like multiple artificial insemination by preferred donors of high genetic quality, as Professor Muller emphasized a quarter of a century ago, and I re-emphasized in my recent Galton Lecture. Such a policy will not be easy to execute. However, I confidently look forward to a time when eugenic improvement will become one of the major aims of mankind.

The motivation for eugenics had been various but now Joshua Lederberg, the geneticist and Nobel Prizewinner of just a few years before, had a new one, more dramatic and urgent than anything before – survival in the short-term:

> In answer to Dr. Bronowski's question about our motivation, I think that most of us here believe that the present population of the world is not intelligent enough to keep itself from being blown up, and we would like to make some provision for the future so that it will have a slightly better chance of avoiding this particular contingency. I am

## GENES, FLIES, BOMBS...

not saying that our measures will be effective, but I think this is our motivation; it is not the negative but the positive aspects of genetic control that we are dealing with here.

So, a century after Noyes and Galton, with a few desperate cries, eugenics slipped away back into the night.

# 11   A CHEMICAL GENETICS

## MACROMOLECULES[1]

Structural organic chemistry began with Kekulé's theory of the carbon-carbon bond in 1858. It opened the possibility that there might be very large, stable organic molecules, such as proteins, that might be important to life. However, with the discovery that some proteins could be obtained in crystalline form and the rise of physical chemistry, the idea of such macromolecules went out of favour and most chemists preferred an explanation of biochemical phenomena in terms of of colloidal or aggregate theory. This abandoned the idea of macromolecules and polymers constructed from strong covalent bonds and instead saw proteins, enzymes and nuclei acids as looser aggregates of smaller molecules. This colloidal theory dominated biochemistry until the late 1920s.

However after that the macromolecule returned to favour as techniques for measuring the molecular weight of proteins, based on the ultracentrifuge and flow bi-refringence, showed that these were large single molecules with well-defined molecular weight. Other substances – including the nucleic acids – showed themselves to be macromolecules.

## X-RAY CRYSTALLOGRAPHY

In 1912 von Laue persuaded two brilliant experimentalists to conduct an experiment that was seriously misconceived but proved to be a milestone in physics. Laue ( the "von" came later) had the idea that the characteristic x-rays emitted by atoms in a crystal, when they were excited by x-rays, might give rise to diffraction patterns because the spacing of the atoms was close to the wavelength of x-rays. When the experiment was conducted and an intense beam of x-rays was directed onto a crystal of copper sulphate and then fell on a photographic plate, it was found that there was a pattern of spots, not dissimilar to the diffraction patterns seen with light.

The effect proved to be nothing to do with characteristic x-rays but was a result of diffraction of the incoming x-rays from different layers of atoms in the crystal. William Lawrence Bragg gave the correct analysis later in 1912 with the Bragg Laws and this effectively triggered the science of x-ray crystallography. Bragg and his father, Sir William Henry, applied the

## GENES, FLIES, BOMBS...

analysis to simple crystals such as table salt and diamond, revealing the ionic nature of the bonds in sodium chloride and the covalent bond of carbon.

The first measurements on organic crystals (anthracene, naphthalene and others) were made by Bragg senior in 1921. While it was not yet possible to determine the detailed structure of these, this did give the intermolecular spacings and possible orientations of the individual molecules. During the 1920s and 1930s experimental and analysis techniques advanced so that the complete structures of large numbers of simpler organic compounds could be determined. Diffraction pictures of more complex structures such as proteins and the tobacco mosaic virus were obtained but they proved impossible to interpret largely because of the phenomenal amount of computation required.

The structure of proteins had long been seen as a particular prize. William Astbury at Leeds had in the early 1930s, in studying fibrous substances like the keratin that makes up wool, concluded that they were composed of long-chain molecules. The diffraction patterns changed when the fibres were stretched suggesting that the molecules were normally folded in some way or coiled. The diffraction patterns were not as distinct as those obtained with crystals but they indicated a characteristic repeat of 510 pm for the un-stretched structures.

By the late 1940s two British groups (Bragg's at Cambridge and Astbury's at Leeds) and that of Linus Pauling in the USA were trying to determine the structure. However their approached were quite different: the British sought the structure by direct determination from x-ray diffraction while Pauling placed great weight on modelling. While ill and bored in 1948 Pauling made a paper model in the helix form using what was known of the components of the protein keratin. He found that, using his knowledge of chemistry, he could make a model of a structure that would probably be stable. There was a problem though: the structure had a repetition distance of 550 pm rather than 510 pm.

Distracted by other matters he did not return to the problem until 1951 when , with his co-workers, he managed to construct a model, called the "α-helix", that seemed good enough to publish and this was done. It still had the flaw of the repetition distance that was too large – but the apparent discrepancy was seen to be insignificant when other keratin molecules showed different distances in diffraction studies. His model was quickly confirmed by experiments by one of Bragg's team.

This was immense progress. The alpha helix proved to be a widespread component of proteins but it turned out to be just a secondary structure.

# A CHEMICAL GENETICS

The proteins themselves were composed of other structures than the helices and the helical regions themselves coiled around one another and folded into complex structures. The proteins would hold their secrets for many more years yet. But the techniques used in finding the α-helix would lead rather quickly to an astonishing leap in our understanding of genetics.

## THE ROLE OF GENES

Bridges's work on inter-sexes in *Drosophila* in the early 1920s had shown that the sex characteristics *Drosophila* were influenced by the balance between the number of X chromosomes and autosomes. He believed that this balance was an essential element in determining other characteristics too. It showed that there was some kind of quantitative relation between the genes and their products. The genic balance idea in sex determination seems to have occurred to Muller in 1912[2]. However, Bridges provided the experimental evidence for it in 1922 – without giving credit to Muller for the idea – that established it.

Garrod's 1902 one gene-one enzyme theory, elaborated by Beadle and Tatum in 1941 after experiments with *Drosophila* and the fungus *Neurospora*,[3] linked the genes with the biochemical processes. It was a concept that showed how the genotype could be isolated from the phenotype: it made any form of Lamarckism even more difficult to sustain scientifically.

George Beadle's review of the field in 1939 saw genetics moving in direction the directions of physiology and biochemistry

> While recognizing that all of genetics must have a physiological basis—and ultimately a physico-chemical one—geneticists for many years have been concerned largely with the numerical and geometrical aspects of heredity, Within recent years, however there has been an evident and growing interest in problems that have to do with the relation of hereditary mechanisms and hereditary units to other branches of biology. A surprisingly large number of specific physiological reactions are now known to be under the control of genes.... The geneticist is, in fact, confident that all processes characteristic of living organisms are ultimately gene-controlled.[4]

At the VIIth Congress Beadle appealed for a breakdown in the barriers between genetics and chemistry. Geneticists had something to offer to chemists he said (although without specifying just what) while chemistry had a great deal to offer genetics.

# GENES, FLIES, BOMBS...

## SCHRODINGER'S BOOK

Erwin Schrodinger was the founder of wave mechanics in the 1920s, developing his famous wave equation over just a few months in Zurich in 1925. He moved to Berlin in 1927 but in the early 30s his distaste for fascism led him to Oxford, to Austria (in an ill-judged move) and then, with the *Anschluss*, to Dublin in 1939. He stayed there until 1957.

He was required to give a series of public lectures and in 1943 chose as his subject the nature of life from a physicist's perspective. The lectures were the source of his book, published the following year as *What is Life?*[5]

The most interesting theme in the book was just how the genetic information could be passed between generations; accepting the existence of genes what could their physical form be? He was aware of the work of Delbruck and the *Three Man Paper* and found the idea that information could be reliably stored in volumes that contained as few as 1000 atoms (as the paper implied) a difficult one. However he concluded that this was possible in a structure – bound by strong chemical bonds – that might be part of an "aperiodic crystal". This would be stable – so reliable – and could carry a "code-script" through the arrangement of individual atoms that could transfer the necessary information.

The influence of the small book on the development of genetics is hotly debated to this day. Francis Crick has said that it stimulated his interest in molecular genetics – he was interested in birds before that – but Max Perutz seems dismissive in a 1987 review : "...a close study of the book and of the related literature has shown me that what was true in his book was not original and most of what was original was known not to be true even when it was written."[6] He had in mind, no doubt and for example, Delbruck's comment that "the gene is a polymer that arises by the repetition of identical atomic structures." However, he did recognise that the book had brought outstanding young physicists into molecular biology and that it had drawn attention to the *Three Man Paper*.

Muller wrote a review[7] soon after the book's publication. He was not impressed by the breadth of Schrodinger's knowledge of genetics (Muller's contribution is not mentioned in the book) but he did concede that the book could (as it did) strengthen links between biology, physics and chemistry. His main criticism was of the Epilogue where Schrodinger becomes more philosophical than scientific and veered off into what Muller called mysticism.

## DNA

Johannes Miescher in 1869 extracted an unusual substance from the

## A CHEMICAL GENETICS

nuclei of cells. It was unusual because, unlike most other biochemical compounds known at the time, it was phosphate-rich and contained nitrogen but no sulphur. He called it nuclein – later changed to nucleic acid. Thirty years later Ludwig Kossel showed that it contained five components: adenine, guanine, thymine, cytosine and uracil – A, T, G, C and U. They appeared to be present in very similar proportions so in 1909 Phoebus Levene put forward what became known as the tetranucleotide hypothesis for one of the nucleic acids – DNA. It proposed that DNA was composed of a string of identical units each with precisely the same structure and with one each of A, T, G and C. The consequence of this was that DNA was subsequently repeatedly dismissed as a possible carrier of genetic information – it was too simple. This was not effectively challenged for more than 40 years.

Three important papers were presented at the VIIth Congress in Edinburgh in 1939. Lewis Stadler showed that mutations were induced in maize pollen most effectively by radiation of a wavelength corresponding to the maximum absorption wavelength of nucleic acid and similar results were reported by Alexander Hollaender for spores of a fungus and E Knapp and H Schreiber for sperm bodies of a liverwort. The nucleic acid, rather than any protein associated with it, began to seem the best candidate for the genetic material [8]

In 1928 Griffith had made an intriguing discovery while investigating the virulence of the *Pneumococcus* bacterium. Its ability to cause infection was related to the presence of a smooth outer covering; the covering became rough if the bacteria were grown for several generations on fresh media and such bacteria (the R strain) did not infect mice inoculated with them. He also found that he could kill the smooth (S strain) so that they lost the power to infect. However, if he mixed the heat-killed S and the R strain bacteria, the mixture regained its virulence. It was as if the R strain was converted back to virulence by something in the dead S strain – something that became known as the transforming factor or transforming principle

The experiments were repeated and the phenomenon was found in other systems but it was not until 1944 that Oswald Avery and his co-workers at the Rockefeller Institute, after a long, complex and careful series of experiments, suggested that the transforming factor was most likely DNA.

Among early comments from geneticists, Muller's was the closest to the mark:

> ... the most probable interpretation of these virus and *Pneumococcus* results then becomes that of actual entrance of the

# GENES, FLIES, BOMBS...

foreign genetic material already there, by a process essentially of the type of crossing-over, though on a more minute scale.[9]

In 1952 Alfred Hershey and Martha Chase worked with bacteriophages, ultra-microscopic packages that attack bacteria. They take over a bacterium's metabolic system to multiply themselves before destroying their host and moving on to infect others bacteria. Phages are composed of two distinct elements: a tail made of protein and a head, a protein shell packed with DNA. When the phage attacks, its tail attaches to and penetrates the bacterium (in this case *E coli*) and injects its own genetic material into the bacteria to hijack it. The question was: is this protein or DNA? By tagging the protein and DNA with different radioactive tracers Hershey and Chase found that the injected material was DNA and they concluded, rather cautiously, that:

> This protein probably has no function in the growth of intracellular phage. The DNA has some function. Further chemical inferences should not be drawn from the experiments presented.[10]

As evidence pointed towards DNA as the carrier of genetic information, studies of the material, based on improved samples and techniques, showed it to be more interesting than Levene's tetranucleotide admitted and that it had the potential to carry the required vast amounts of information. In 1950 Erwin Chargaff and his co-workers[11] found that that the four key components (adenine, guanine, thymine and cytosine A, T, G and C) were present in closely related proportions. The amount of A was always the same as the T; the G the same as C but A and T were present in significantly higher proportions than were C and G. At a stroke this threw out the tetranucleotide model and opened up the possibility of DNA as an information carrier. The fact that the proportions varied between species just added to DNA's interest.

The realisation that DNA was a double helix came in 1953 to Francis Crick and James Watson. Starting in 1951 they took x-ray diffraction patterns obtained by two other researchers – Rosalind Franklin and Maurice Wilkins – and constructed cardboard models using what they knew of the composition and bonding of the components. In this they consciously followed the method that had led Linus Pauling to the α-helix – and were stimulated to greater efforts when Pauling came up with his own (an erroneous triple helix) DNA model. Finally on 28 February 1953 the endless model building paid off when they found a structure that met all the requirements of stability and composition. It was a double helix with the remarkable properties that it could carry the genetic information and replicate itself. Not quite an instant sensation, it was soon recognised as a turning point in genetics.

## A CHEMICAL GENETICS

Controversy remains over the use of Rosalind Franklin's diffraction photographs one of which clearly showed, to those immersed in the problem as Crick and Watson were, that the molecule had a helical form. The problem was that Crick and Watson were shown the photographs without Franklin's knowledge. Franklin received hardly any acknowledgement for her brilliant and crucial experimental work. Crick, Watson and Wilkins shared a Nobel Prize in 1962. Rosalind Franklin died of ovarian cancer in 1958 having made friends of Watson and Crick.

It remained to explain quite how the replication worked and how DNA actually directed the manufacture of proteins. The replication process became clear within a few years largely as a result of the work of Matthew Meselson and Franklin Stahl in 1958. In 1957 Francis Crick could outline the involvement of RNA as an intermediary in the creation of proteins based on DNA as a template and allowed him to state what he called the Central Dogma of genetics: "...once 'information' has passed into a protein *it cannot get out again.*"[12] Intense activity over the next few years revealed the many secrets of the processes. A problem that remained unsolved for some time was quite how the four digit code of DNA was converted into specifying the twenty amino acids strung together to make a protein chain. Crick speculated about this in 1957 – there had to be a code of some kind – but the code remained un-cracked until the mid-1960s and the work of Nirenberg, Khorana and Holley.[13]

It was, of course, just the beginnings of the extraordinary science that is molecular genetics. There was so much more to discover about how genes were regulated and the details of protein synthesis, DNA was to be sequenced and then to be engineered. However the mid-1960s marked the end of the beginning, where the major ideas and principles had been set down. Muller lived to see, and even have a small part in, this great illuminating breakthrough.

# 12  POSTSCRIPT

## *MULLER – HEALTH*

While he was in Hawaii, in October 1953, Muller had a health check-up. It revealed that he had high blood pressure and, more alarmingly, that there was evidence of a past heart attack. Muller described the result as "coronary insufficiency"; it would cut down his outdoor activities, probably for ever. He kept it from most of his co-workers and students. He returned to Bloomington and resumed his six and a half day week. The concessions to his condition were a nap each afternoon and a daily walk to and from the lab.

Illustration 10: Muller in 1956

# POSTSCRIPT

Over the next few years he travelled quite extensively. To Europe in 1955, Nova Scotia in 1957 for the Pugwash conference on disarmament. His work at Bloomington hardly slackened.

When he was 70 in 1960 there was a banquet in his honour at the Columbia University Faculty Club in New York. There were speeches of course and Altenburg's was so effusive that Muller seemed to bow his head in embarrassment. When he suddenly slumped back in his chair and passed out people realised it was rather mores serious than that. He was laid out on the floor, his clothing loosened and a doctor was called. He quickly recovered – he had earlier taken a double dose of his blood pressure medicine by mistake. When the doctor arrived Altenburg could introduce the the patient with the words "May I introduce you to Professor Hermann Muller, discoverer of dosage compensation!" Muller was kept in hospital over-night and was disappointed that Thea had cancelled the talk he was to give to the Evolution Society at the AAAS the following day.

He should have retired at 70 but managed to persuade the University that he should carry on. This he did until 1964. He was actively promoting the teaching of evolution in schools, still involved in radiation protection standards with the NCRP and a supporter of the fallout shelter programme. He was promoting a negotiated test ban treaty. In 1963 the American Humanist Society named him Humanist of the Year.

After retirement, finally, in 1964 he took a position at the City of Hope Medical Center in Duarte, California. He expected to spend some time there, at their Center for Advanced Study in Medicine and the Life Sciences, writing. A priority was to be launching his Foundation for Germinal Choice. He had just begun the prearations for this when he was struck by a severe heart attack. Even after a lengthy recuperation he was frailer and less energetic than his old self. The eugenics plans slipped.

His mind was still clear and sharp and when he and Thea moved to Madison, Wisconsin for its clean air in the fall of 1965 he enjoyed teaching a course on evolution at the university. He wrapped up and took daily walks on which he was hardly ever really warm. His old eye condition returned.

They moved back to Bloomington in the summer of 1966, into a new house where he didn't have to manage stairs. He delivered a speech in Chicago to the Third International Congress on Human Genetics in September 1966. In *What Genetic Course Will Man Steer?* he promoted his ideas for germinal choice. It was to be his final speech.

The heart condition worsened and on his $76^{th}$ birthday, 21 December 1966, he was taken to hospital suffering from kidney failure. It was the

# GENES, FLIES, BOMBS...

beginning of the end and over the next few months, through periods of confusion and dismay, Muller realised it. Rather than linger on he stopped eating and, on 5 April 1967, he died of congestive heart failure.[1]

## MULLER'S CONTRIBUTION

What was indisputable was that he devised the careful experiment that revealed radiation caused mutations and quickly became a tool that allowed the generation of vast number mutants that permitted mapping of fly genes using the clever insight of Sturtevant. That established the chromosome as a string of genes as a reality for most scientists. It won the Nobel Prize for him. Stadler was of course close behind, delayed only by a less amenable experimental subject than Muller's flies.

Long after that happened there continued to be doubts about exactly what mutations were (and Muller shared some of them for a while). Perhaps invisible rearrangements of parts of the chromosome, perhaps real changes in content. It would take molecular biology and DNA to clarify that.

He was an idealist in several senses of the word. In his political life – certainly up to World War 2 – he hoped for the Communist dream, a reverie of equality and progress. His life-long belief in eugenics drew on the same energy source. In both cases he was disappointed. His belief in Communism was crushed by the realisation of how it had turned out in practice and by Stalin's perverted wrath. His support for eugenics lasted as long as he did but by the end it had become an unpopular, withered and even derided belief in the power of the sperm of great men.

Idealism of a different kind pervaded his work. He believed, following the rest of the Morgan school, that the normal chromosome (the wild-type) varied very little between members of a species. Mutations were very rare variations of this ideal. This belief, that harked back to much earlier ideas of nature, proved to be wide of the mark. As Sewall Wright claimed in the disputes at the BEAR Committee, there were many neutral mutations and many variations that had no effect on viability. Muller's conviction led him to the genetic death idea which proved both graphic and persuasive – but which was ultimately mistaken.

In some sense this did not matter because it was simply an expression of his conviction that radiation was damaging and bad for you, a generally unpopular assertion when he made it in the 1920s to radiologists and not welcome when the AEC were developing better bombs in the 1940s and 50s. The principle Muller established that even low levels of radiation could cause genetic defects was hardly challenged and it might have been extended to somatic effects earlier than it was. But that had to wait for

# POSTSCRIPT

incontrovertible human evidence after the bombs.

Of course, it is possible that the overstatement of genetic hazards (something hardly surprising when the population of much of the world's population appeared threatened by fallout) somehow obscured the somatic risks of radiation. But we can hardly blame Muller for that.

Personally many found him difficult to work with and saw him as obsessed not just by his science but by priority and recognition. He was not always regarded as a good teacher but with right people, bright and committed, he was an inspirational co-worker and tutor for whom they had great respect. He was admired – and still is – by many more as a maverick , an anti-establishment figure whose persecution by the AEC stands as a reminder of the fragile nature of the link between science and government. For some though he remains an enigmatic figure, a brilliant scientist who was seduced by quasi-scientific social ideologies for much of his life.

Whatever you think of Muller and his contribution it is clear enough that the star of the story is genetics itself. After the rediscovery of Mendel in 1900 and the growing understanding of the cell and its components that took place, the association of heredity with the chromosomes was a major insight. The adoption of the small fly as an experimental material by Morgan and his group was one of the momentous choices in science, a serendipitous and clever selection that led to two or three golden decades in genetics. Sturtevant's insight that it would be possible to map the genes along the chromosome may have come overnight but it was to have a profound and lasting impact. Spurred on by Muller's discovery that mutations could be created almost at will and used for this mapping, Morgan's small group, including Calvin Bridges, virtually invented genetics. When this flood of ideas from Morgan's people and soon from others began to slow in the 1940s, there was already excitement as chemists and physicists took an interest. Soon enough biochemistry and DNA came along and a molecular biology gave deeper insights into heredity, disease and evolution – something Muller strived and hoped for.

# 13  ENDNOTES

## 1 GENESIS

1 *Loimographia: An Account of the Great Plague of London in the Year 1655*, ed J F Payne London: Epidemiological Soc London 1894(Wilson C 1995)

2 William Harvey believed that Aristotle had studied no more than three eggs.

3 (Mayr E 1982) p636

4 (Mayr E 1982) p637

5 See Ruestow on the microscope(Ruestow E G 1996)

6 Although, of course, the image quality and convenience of the modern instrument are far better.

7 (Moll W 2006).

8 *Bombyx mori*, the cultivated silkworm, had been studied previously because of its economic importance but Malpighi was the first to subject it (or any other insect for that matter) to detailed microscopic dissection. The book contained 48 drawings. The Royal Society's copy has an inserted watercolour of the larva thought to be by Malpighi. (Cobb M 2000) (Cobb M 2002)

9 See p135 (Wilson C 1995)

10 (Cobb M 2000)

11 An anagram of the Latin version of Plantades' name

12 See (Cole F J 1930)

13 For an interesting short history of many of the issues in this section see (Clarke G N 2006)

14 From Baker H, *The Microscope Made Easy,* 2nd ed 1743 p252 as quoted by (Woodruff L L 1921)

15 (Benson K R 1991).

16 Ibid

17 (Olby R 1985)

18 Ibid

19 (Lovejoy A O 1936)

20 With a careful gradation from man to the animals via negroes and wild children to apes

21 (Olby R 1985)p2

22 *Systema Naturae* covered the classification of both animals and plants and extended to 13 editions if we include the edition edited and amplified by J F Gmelin published around 1790. The book, which originally had nine pages, had grown to several thousands by then.

# ENDNOTES

23 An alternative story has Linnaeus naming the weed before the falling out. The slight came when Siegesbeck, after his attack, found some seeds (or fruits) of the weed in an envelope (or casket) with "cuculus ingratus" (ungrateful cuckoo) written on it in Linnaeus's hand. It may not even be a useless weed : in 2009 its extract was being marketed "to relieve rheumatic conditions, to improve the motility of joints, and to counteract toxicity."

24 Linnaeus himself thought that, while his classification of animals represented some real biological distinctions, his system for plants simply helped in identification.

25 See Hybrids in the Glossary

26 See (Browne J 1989) for a summary.

27 (Olby R 1985) p4

28 Ibid p21

29 (Sachs J von 1890) p425

30 Seemingly because he gave a poor answer to a question on the classification and use of the *Mammalia* (Olby p95)

31 In an analysis of the light expected from double stars.

32 (Olby R 1985) . But some (Mayr E 1982) have seen an intention to determine the quantitative pattern of heredity.

33 (Mendel G 1866) A translation is available on www.mendelweb.org

34 Although it has been argued, notably by R A Fisher, that Mendel's results from which he deduced the factor were better than could reasonably have been expected. Mendel, Fisher thought, was selecting his data to support an idea he already considered proven.

35 Some of the conflicting ideas about Mendel's work are discussed by Olby (Olby R C 1997),Sapp (Sapp J 1990) to name but two.

36 It is often suggested that Mendel was prompted to try *Hieracium* as a subject by Karl Nägeli but there is evidence that Mendel had chosen it before entering into correspondence with Nägeli. There is also some dispute about whether Mendel reacted with despair to the differences or found them interesting.

37 (Olby R 1985)

38 (Gliboff S 1999)

## 2 EVOLUTION, CELLS AND CHROMOSOMES

1 Both quotations from autobiographical notes prepared by Muller. See (Carlson E A 1981) p15

2 In the late 1700s, James Ussher, the Bishop of Armagh, calculated the precise date as Monday 23 October 4004BC – at 9 AM. The 4004 BC date appeared in the marginal commentary of Bibles of the time.

3 p307 (Mayr E 1982)

4(Winchester S 2002)

# GENES, FLIES, BOMBS...

5 A conflict between European powers that started in 1756 and ended in 1763. Britain and Prussia with support from some small German states fought France, Austria, Russia, Saxony and Sweden. Portugal and Spain joined in later while Russia briefly switched sides. Fighting took place on a global scale, notably in North America, leading Winston Churchill to describe it as the first World War.

6 (Magner L N 1979)

7 Although Mayr argues otherwise

8 (Anon (Chambers R) 1844)

9 (Mayr E 1982).

10 Ibid p393

11 (Darwin C R 1868)

12 Ibid p396

13 Ibid Author's Preface

14 (Mayr E 1982)

15 *Unsere Körperform und das physiologische Problem ihrer Entstehung*, Vogel, Leipzig.

16 The remains were exhumed from the churchyard of the Johanneskirche, Leipzig in 1894 before the church was extended. They were reburied inside the church but moved to the Thomaskirche in the city in 1950 after the Johanneskirche was badly damaged in WW2. The identification, it has to be said, is disputed to this day.

17 (Weismann A 1889)

18 See Darden p82(Darden L 1991)(Mayr E 1985)

19 (Harris H 1998)

20 Ibid For more details including the effect of Remak's Jewishness on his career. There have been accusations of plagiarism against Virchow. He does seem, at the very least, to have been ungenerous in his acknowledgement of what Remak had achieved

21 (Mazzarello P 1999)

22 Waldeyer introduced the term "chromosome" in 1888 and Flemming's term came to mean something slightly different.

23 (Mayr E 1982) p67)

24 (Balzer F 1967)

25 (Maderspacher F 2008)

26 (Sutton W S 1902)

27 (Sutton W S 1903)

28 (Crow E W and Crow J F 2002)

29 (Mayr E 1982) p734

30 Catherine Auerbach has argued that even these were not terminal and that the theory lived on until the molecular biology dealt the *coup de grace*.

# ENDNOTES

*(Auerbach C 1976)*

31 (Timofeeff-Ressovsky N , Zimmer K G and Delbruck M 1935)See the discussion in Chapter 4.

32 Darwin did not accept Galton's experiments as the death knell for pangenesis arguing that he had never suggested that the gemmules circulated in the blood – how would that explain heredity in the bloodless Protozoa for example? Darwin claimed they "diffused". A summary of the dispute and Darwin's continuing reluctance to abandon pangenesis can be found in (Robinson G 1979) Chapter 1.

33 See Gillham (Gillham N W 2001)

## 3 FLIES, GENES AND MUTATIONS

1 http://www.wymanpiano.com/altenburg/history.html

2 His three volume *Orthogenetic Evolution in Pigeons* was published posthumously in 1919. He was active in efforts to save the American Passenger Pigeon from extinction but the last of these birds, flocks of which had once darkened the skies for days on end as they passed overhead, died in Cincinnati Zoo in 1914.

3 He had visited de Vries's garden and this may have inspired him to change from embryology and development to genetics.

4 (Morgan T H 1910a)

5 *The Mechanism of Mendelian Heredity* (Morgan T H, Sturtevant A H , Muller H J and Bridges C B 1915)

6 (Morgan T H 1934) For more about Morgan see (Sturtevant A H 1959) and (Sturtevant A H 2001)

7 The related *D. bifurca* has an even longer sperm which reaches 6 cm. This is about 1000 times longer than typical human sperm and the longest known in the animal kingdom. The extraordinary length, which is almost entirely tail, is thought to be the result of sperm competition in the female's reproductive tract – which is not much shorter than the uncoiled sperm's length.

8 (Sturtevant A H 1959)

9 Some historians rather doubt that de Vries anticipated Boveri, believing that he was talking about something rather different. However, Boveri himself thought that de Vries was there first. (Darden L 1991)

10 (Morgan T H and Cattell E 1912)   (Morgan T H 1910b)

11 (Sturtevant A H 1913)

12 Crossing over had been found generally to be less in females than males in most species but the absence in the male, as in *Drosophila*, is very unusual and later proved to be another useful feature of the fly. It comes from the failure of the chromosomes to stick together in prophase I of meiosis but quite why this happens seems still a mystery

13 (Morgan T H, Sturtevant A H , Muller H J and Bridges C B 1915)

14 Quoted in (Sturtevant A H 1965) p49

15 These quotes taken from (Cock A G 1983)

16 Ibid

17 According to Carlson it was an idea that Muller had in 1911 (Carlson E A 1981) p79

18 (Altenburg E and Muller H J 1920)

19 (Carlson E A 1981) and (Dunn L C 1965)

20 The controversy was worked out in (Castle W E 1919a; Sturtevant A H, Bridges C B and Morgan T H 1919; Castle W E 1919b)

21 (Dunn L C 1965)

22 (Bateson W, Saunders E R and Punnett R C 1904)

23 (Bateson W 1909). The situation became rather confused when R A Fisher used the term "epistacy" – which quickly transformed through use into "epistasis" – to describe more general gene interactions (Fisher R A 1918). People were still trying to straighten things out at the end of the 20$^{th}$ century. See for example (Phillips P C 1998) and the suggestion that it might help if the term "gene interactions" were used.

24 Bridges' personal papers were destroyed by Morgan and Sturtevant after his death in an attempt to draw a veil across his colourful sex life (Private communication, de Jong Lambert 2015)

25 (Sturtevant A H 1965) p54

26 See Glossary

27 (Carlson E A 1981)

## 4 MUTATIONS FLOURISH

1 Several of Morgan's staff had been drafted into the services. Sturtevant was serving six months as an autopsy assistant in 1918 and 1919. Muller's hopes of a permanent appointment were dashed by Wilson – Muller always mistakenly blamed Morgan.

2 "Inhibitor" is a misnomer. The crossovers take place but the resulting progeny are not viable and abort

3 St Petersburg was renamed Petrograd in 1914, became Leningrad in 1924 and then returned to St Petersburg in 1991.

4 There is an account of some aspects of the visit to Russia in (Muller H J 1923)

5 The paper is (Muller H J and Jacobs-Muller J M 1925). More details of Jessie's career can be found in (Green J and LaDuke J 2009)The information about the honeymoon comes from (Carlson E A 1981) p133. Carlson has a different date for their marriage than quoted here.

6 The twins were reported 1922 by Paul Popenoe. Popenhoe features in the chapter on eugenics

7 (Muller H J 1925; Muller H J 1926)

# ENDNOTES

8 (Carlson E A 1981) p141

9 Ibid p143

10 Ibid p146

11 (Muller H J 1927)

12 (Carlson E A 1981)p148

13 (Muller H J 1928)

14 (Weinstein A 1928) , (Whiting P W 1928)

15 Apologies to Marxian scholars for this travesty of an account of the clash of philosophies – but something had to be written.

16 (Carlson E A 1981) p175.

17 He also seems to have been involved in Communist activities more generally although he claimed never to have been a Party member. Ibid footnote p166

18 (Carlson E A 2011)

19 The files eventually released were extensively redacted with some 80% of the text being obliterated. They throw little light on Muller's activities.

20 See Glossary

21 (Carlson E A 1981) p182

22 Renamed the *Steuben* the ship was used as a troop carrier during the war. She was sunk in 1945 by a Soviet submarine with the loss of more than 3000 lives. Some others were not so lucky – torpedoed (Crow J F 1992)

23 (Sturtevant A H 1925)

24 (Bridges C B 1936)

25 (East E M 1936)

26 (Shull G H 1948)

27 (Crow J F 1998)

28 Ibid

29 (Springer N M and Stupar R M 2007)

## *5 EVOLUTION AND EXILE*

1 See (Edwards A W F 2008) for details of Hardy's involvement

2 (Provine W B 1971)

3 Some butterflies, palatable to predators, are patterned similarly to toxic species to make them less attractive as a food source. Bateson suggested an evolutionary mechanism for this and it became known as Batesian mimicry.

4 See *Population Genetics* by M B Hamilton

5 See (Sarkar S 1992) for a review of Haldane's contribution.

6 (Haldane J B S 1924)

7 See (Grant B S 1999) for a review and account of some of the problems associated with Kettlewell's work.

# GENES, FLIES, BOMBS...

8 Wright was the offspring of first cousins.

9 Fisher thought extremely low selection pressure, given enough time could produce evolutionary change but Wright emphasised other processes under those circumstances such as genetic drift.(Bourguet D 1999)

10 See (Bourguet D 1999) and (Crow J F 2008)

11 (Barker M and Lerner I M 1961)

12 See (Mayr E and Provine W B (eds) 1980) pp242-278 and (Adams M 1968)

13 (Adams M 1968)

14 (Dobzhansky T 1937)

15 (Bohr N 1933)

16 (Carlson E A 1981) p188

17 See (McKaughan D J 2005) for Delbruck's motivation.

18 (Timofeeff-Ressovsky N , Zimmer K G and Delbruck M 1935). A translation into English can be found in (Sloan P R and Fogel B 2011). The iconic paper is popularly known as the *Three Man Paper* or the *Green Pamphlet* (from the colour of its cover).

19 From the translation in (Sloan P R and Fogel B 2011).

20 Sometimes translated as Cultivation or Breeding rather than Industry. Carlson (Carlson E A 1981) refers to it as the Institute of Applied Botany. It is now known as the N. I. Vavilov Research Institute for Plant Industry.

21 (Carlson E A 1981)

22 See (Graham L R 1993)pp235-244

23 See (Carlson E A 1981) p204 and 212. Robert Miller and Jenny returned to the USA (via Spain) in 1939. He was investigated as a suspected Soviet spy by the FBI in the 1940s

24 Published by Izdatel'stvo Akademii Nauk SSR in Moscow and Leningrad in 1934. The piece is most easily accessible as Appendix II of Loren Graham's 1972 book Science and Philosophy in the Soviet Union.(Graham L R 1972) It is not reproduced in the follow-up volume of 1987.

25 (Michurin I V 1934)

26 (Joravsky D 1970) p87

27 Ibid

28 Ibid p214 from *Novoe v nauke* 1952

29 This has to be seen in the context of the Great Terror (or Purge) in the late 1930s when some one million people (and maybe more) were executed or died as result of their treatment in prison camps.

30 Three speeches from the 1939 conference illustrate the differences rather well (Vavilov, Lysenko and Polyakov 1940)

31 Source: Preface by P N Yakolev in *I. V. Michurin: Selected Works*, Foreign Languages Publishing House, 1949

32 (Graham L R 1987)

# ENDNOTES

33 The literature associated with Lysenko is enormous. A few examples with different viewpoints are: (Roll-Hansen N 2005)

34 (Muller H J 1936)

35 (Muller H J 1948b) p369. He continued the attack on Lysenko and USSR at the 8th International Congress on Genetics(Muller H J 1949), in the Bulletin of the Atom Scientists (Muller H J 1948b) and elsewhere(Muller H J 1951).

36 (Joravsky D 1970) See also (Paul D B 1983)

37 (Glad J 2003)

38 Perhaps more by the clear statement of non-Lysenkoist ideas than by the eugenic content.

39 Letter from Muller to Huxley (Carlson E A 1981) p233

40 Some details can be found in (Palfreeman L 2015).

41 (Bennett M & Grunfeld A T (ed) 2004

42 (Preston P 2012)

43 Ibid

44 The knowledge about the relationship comes from Milly Bennett's papers in the Hoover Institution archives at Stanford University (Kirschenbaum L A 2015)

45 de Jong-Lambert has suggested that Milly was one of a number of women Muller was involved with in Russia, taking advantage of his status as a celebrated scientist (Private communication)

46 For some general background see (Falconer D 1993) and for the budgerigar work see the book by Crew and Rowena Lamy (Crew F A E and Lamy R 1935).

47 From Crew's recollections of the early years of the Genetics Society and the beginnings of *Drosophila* research in the UK in (Crew F A E 1969)

48 (Muller H J and MacKenzie K 1939)

49 (MacKenzie K and Muller H J 1940)

50 He wrote *The remaking of the chromosomes* while he was there (Muller H J 1938)

51 Included in (Muller H J 1939) pp14-15

52 Russ's actual words were " In the absence of proof it would greatly strengthen the argument if a suitable proviso were inserted before the question of applicability to man was discussed." See (Carlson E A 1981) p255

53 (Muller H J 1940)

54 (Ray-Chaudhuri S P 1939)

55 (Punnett R C 1941) See "R C"(R C 1941) for a brief summary of the events at the congress and a review that is more passionate than many.

56 The service had been inaugurated in May 1939. The crossing from New York took about 26 hours in total, including a refuelling stop at Horta in the Azores. The westward flight usually took longer because of prevailing westerly winds. The service continued throughout the war.

# GENES, FLIES, BOMBS...

57 Taken from (Carlson E A 1981) p272-3 who drew from a conversation with Thea in 1973.

58 (Provine W B 1980). Provine was actually considering contributions relevant to evolutionary thought but this seemed to encompass most of what was going on.

59 (Goldschmidt R B 1946)

60 (Hilbert D 1930) As quoted (and translated?) by Joravsky in (Joravsky D 1970) p202

61 (Stadler L J 1932)

62 (Rhoades M M 1957).

63 (Muller H J 1940)This paper discussed the different break/dosage regimes and theories.

64 (Carlson E A 1981) p209

65 The Fisher and Muller references are (Fisher R A 1930)and (Muller H J 1932).Crow's summary comes from (Crow J F 2006).

66 (Muller H J 1958)

67 (Muller H J 1964c)

68 (Felsenstein J 1974)

69 .(Muller H J 1964a)

70 (Muller H J 1943)

71 The information on Raffel is based on obituaries in local newspapers, US government sources such as Censuses, the memoirs of T M Sonneborn (Sonneborn T M 1978) and comments by Muller reported by Carlson(Carlson E A 1981)

72 (Stein G 1937). The information about her death comes from Anna Linzie's book (Linzie A 2006) p197

## 6 BEFORE THE BOMB

1 (Coppes-Zantinga A R and Coppes M J 1998)

2 (Mould R F 1995)

3 Ibid

4 (Doll R 1995)

5 (Colwell H A and Russ S 1934)

6 (Martland H S 1929; Martland H S 1931; Martland H S, Conlon P and Knef J P 1925)

7 (Doll R 1995)

8 (Sturtevant A H 1965)

9 (Bergonie J and Tribondeau L 1906)

10 (Bardeen C R 1907)

11 (McGregor J H 1908; Regaud C and Dubreuil G 1908)

12 (Mavor J W 1923; Mavor J W 1922; Mavor J W 1924)

13 (Muller H J 1927)

14 (Hanson F B and Heys F 1929)

15 Published in German in 1914 (Boveri T 1914) and in English in a translation by his wife (Boveri T 1929) in 1929.

16 See (Soto A M and Sonnenschein C 2014) and references there for a history of the theory (and the opposing "tissue theory") and its current relevance.

## 7 THE BOMB

1 This was the conclusion of Francis Perrin.

2 The fallout from the bomb contaminated packaging material made at an Indiana plant for Kodak and spotted photographic stock was returned by customers. A Kodak scientist cited (Webb J H 1949) fallout contamination of the river water used by the plant as the likely source. When similar problems were found following the early Nevada tests, the AEC agreed to issue secret advance information about tests to the photographic industry.

3 (Beebe G W 1979)

4. (Plummer G 1952)

5 (Schull W J 1995)

6 (Folley J H, Borges W and Yamawaki T 1952)

7 (Moloney W C and Kastenbaum M A 1955)

8 (March H C 1944), (M(Anon 1947)arch H C 1950)

9 (Anon 1947)

10 (Neel J V & Schull W J 1956). The claim of an effect on sex ratio in a preliminary report in 1953 was withdrawn in this report.

## 8 FALLOUT

1 (Muller H J 1964b)

2 It used the core that had killed Harry Daghlian in 1945 and Louis Slotin in May 1946 in two criticality accidents.

3 On the same day the *Times* published the report of the British Mission to Japan to study the effects of the bombs.

4 Not to be confused with Shields Warren

5 How useful the information supplied by Fuchs actually was in the development of the Russian H-bomb is still debated. He was present at the Tripartite conference held in London in November 1948 having prepared a paper on Harwell effluent with L H Gray. He made other contributions on dose measurement and internal radiation.

6 (Eisenbud M and Harley J T 1958)

7 Instead of relying on the energy released when nuclei of heavy elements split, fusion devices work by exploiting the energy released when nuclei of light elements fuse. The fission trigger is needed to get this process started.

# GENES, FLIES, BOMBS...

8 The US were unable to release details of the radioactive composition of the fallout because that would have given the USSR some insight into the design of the bomb.

9 The report, issued on 15 February 1955, was reproduced in full in the Manchester Guardian of 1 March (USAEC 1955)(the anniversary of the Castle Bravo test) with a note that "no comparable information has been issued in Britain."

10 The units of the time were roentgens but these have been converted into Grays – with some approximation.

11 Radioactive isotopes of strontium and iodine produced by weapons which accumulate in bone and the thyroid gland respectively.

12 Manchester Guardian 1 April 1954

## 9 RADIATION HARMS

1 See (Meggitt G C 2008) for more detail

2 (IXRPC 1934)

3 See footnote p106 (Jones C G 2005) for the reasons: essentially an increase in the safety margin on the erythema dose and doubts over measurement capabilities.See (Cantril ST and Parker HM 1945) for the reasoning for this and and some of the other problems of principle and practice.

4 (Taylor LS 1990)

5 (Taylor L S 2002)

6 ICRP was created after the war by the International Congress of Radiology. It continued and expanded the role of the International X-ray and Radium Protection Committee that had been set up in 1928.

7 (Taylor L S 1984). A very slightly different version of the Chalk River meeting of 39/30 September 1949 exists(Warren S et al 1949)

8 (ICRP 1951)

9 It is difficult to compare the old and revised doses because they were expressed slightly differently. This is the dose measured if free air but ICRP emphasised the importance of surface dose and defined the new limit (5 mGy per week) in terms of this. There were further limits for beta radiation and for neutrons and limits for partial body exposures and exposure of critical tissues (3 mGy per week at the blood-forming organs).

10 The name comes from the fact that, for example 1 Gray's worth of alpha-particles are equivalent in biological effect to 20 Gray's worth of x-rays. In the USA some older units are still in widespread use. Instead of the Gray, the rad is used with 1 Gy= 100 rad. The unit for the equivalent dose is the rem with 1 Sv = 100 rem.

11 (Muller H J 1950)

12 (Muller H J 1947)

13 (Muller H J 1948a)

14 (Wallace B 1970) p59

# ENDNOTES

15 (BEAR Committee 1956b; BEAR Committee 1956a)

16 (Wallace B 1970), (Wallace B 1991)

17 (Denniston C 1982)p341

18 (Tyzzer E E 1916)

19 (Paigen K 2003a; Paigen K 2003b)

20 (Charles D R 1950) and (Charles D R, Tihen J A, Otis E M and Grobman A B 1961)

21 (Grobman A B 1951). A review by James Neel (Neel J V 1951) and the response by Grobman (Grobman AB 1952) make interesting reading. According to Rader (Rader KA 2004)p242 the book drew little official response except "a dismissive press statement" from the AEC.

22 (Rader K A 2004)

23 More simply, easily recognisable characteristics would be found that resulted only when both copies of a gene were defective. Female mice would be bred to have the two defective genes at seven loci. When mated with wild males with two working genes at a locus all the offspring would have at least one working gene and would not show the characteristic. However if one of the genes of the male was mutated to make it stop functioning then there would be mutant offspring with clearly visible evidence.

24 (Russell W L 1989)

25 See Davis AP and Justice MJ 1998 for further details.

26 (Russell W L 1951)

27 (Russell W L 1954)

28 (Alexander M L 1954)

29 (Russell W L, Russell L B and Kelly E M 1958)

30 (Russell W L, Russell L B, Gower J S and Maddux S C 1958)

31 (Russell WL 1977)

32 (Carter T C, Lyon M F and Phillips R J S 1956)

33 (Lyon M and Morris T 1966)

34 (Haldane J B S 1956)

35 (Carter TC 1959a)

36 Histories of the Mouse House include: (Russell L B 2013); where a fairly complete bibliography of the work performed by the Russells can be found, and (Russell W L 1989).

37 An interesting paper by Soraya de Chadarevian(de Chadarevian, S 2006) reviewing the history of and motivation for the Harwell work reveals this in a footnote on page 782. The paper is slightly marred by a mis-spelling of Mary Lyon's name as Mary "Lyons" throughout.

38 As reproduced in his 1989 Reminiscences(Russell W L 1989).

39 Ibid. Other useful sources are (Russell W L 1956) and (UNSCEAR 1958).

40 Henderson was the third son of the former Labour leader Arthur

# GENES, FLIES, BOMBS...

Henderson.

41 Himsworth held the post for nearly twenty years and was effectively the Chief Executive of the Medical Research Council. He was appointed KCB in 1952, was an Honorary Physician to the Queen and was elected FRS in 1955.

42 (MRC 1956)

43 (Court Brown W M and Doll R 1957)

44 The comments here are based on the summary reports published in the journal *Science*(BEAR Committee 1956a),(BEAR Committee 1956b) and the summary report for the public (BEAR Committee 1956c).

45 (Anon 1960)

46 Crow (Crow J F 1995)says that 10 r (100 mSv) was suggested, for reasons that were unclear even at the time, by Russell.

47 Ibid for the division and (Hamblin J D 2007) for the split.

48 The complete record is available on line in three parts http://catalog.hathitrust.org/Record/001621971 (Anon 1958) together with a summary document (Joint Comm Atomic Energy 1957).

49 (ICRP 1955)

50 The report on rats was from Russ and Scott in 1939; Henshaw found similar in mice. The Shields Warren paper (Warren, Shields 1956) analysis was overturned by Selser and Sartwell (Selser R and Sartwell P E 1958). For general comments on accelerated ageing as a cause see (Mole R H 1959). Egon Lorenz reported the life extension in 1950 (Lorenz E 1950) – a result still quoted in the $21^{st}$ century (Muckerheide J 2000) in support of hormesis – although he did not draw attention to it.

51 The information about the development of the genre comes from *Celluloid Mushroom Clouds* (Evans J A 1998)

52 (Binder O O 1953)

53 Not to be confused with the Miss Atom contests run by the Russian nuclear industry from 2004 to 2011 which aimed to show that employees of the industry could look good.

54 (Harada T and Ishida M 1960).

55 (UNSCEAR 1962) Annexe D p149 for details .

56 (Beebe G W, Ishida M, Jablon S 1962)

57 The United Nations Scientific Committee on the Effects of Atomic Radiation (UNSCEAR)was created in 1955 by the General Assembly of the United Nations, in the throes of concern about fallout, as an authoritative body on the effects of radiation. Its regular reports remain excellent summaries of the field.

58 (Jablon S, Ishida M and Beebe G W 1964)

59 (ICRP 1964) p22

60 The calculations were actually of a quantity called kerma rather than absorbed dose.

61 From (Jablon S and Kato H 1972) and UNSCEAR 1972 Annex H (UNSCEAR

# ENDNOTES

1972). Particularly see fig 16 of Jablon and Kato for the change in number of deaths in the high dose group with time for leukaemia and other malignancies.

62 (UNSCEAR 1972) Table 22 and text.

63 (Kato H and Schull W J 1982)

64 (Kato H, Schull W J and Neel J V 1966)

65 (Neel J V, Kato H and Schull W J 1974)

66 Early results of the survivor studies are described in (Awa A A,Honda T,Sofuni T,Nerrishi S, Yoshida M C and Matsui T 1971). The work on offspring to the mid-70s is summarised in (Awa A A 1975).

67 The BEIR reports (BEIR Committee 1972; BEIR Committee 1979; BEIR Committee 1980; BEIR Committee 1990; BEIR Committee 2006) considered these models in great detail. Various UNSCEAR and ICRP reports have some detail also.

68 (Preston D L, Shimizu Y, Pierce D A Suyama A and Mabuchi K 2003; Preston D L, Shimizu Y, Pierce D A, Suyama A and Mabuchi K 2003)

69 (Ozasa K , Shimizu Y , Suyama A , Kasagi F, Midori Soda M ,Grant E J , Sakata R, Sugiyama H and Kodama K 2012)

70 (Denniston C 1982)

71 The estimates coming from the Japanese bomb studies were not really used in risk estimation. See (Sankaranarayanan K and Chakraborty R 2000c) and(Sankaranarayanan K and Chakraborty R 2000b).

72 (Neel J V 1999)

73 (Haldane J B S 1955)

74 (BEIR Committee 2006)

75 (Chakraborty R N and Yasuda C 1998)

76 (Sankaranarayanan K and Chakraborty R 2000a)

77 This from UNSCEAR Annex A of 2001 (UNSCEAR 2001). Further details and discussion can be found in (Sankaranarayanan K and Chakraborty R 2000), (Sankaranarayanan K and Chakraborty R 2000b), (Sankaranarayanan K and Gentner N E 2002), (Sankaranarayanan K and Chakraborty R 2000c). Some discussion also in BEIRVII Ph2 p125

78 (Sankaranarayanan K 2001)

79 UNSCEAR 2001

80 (Falconer D S 1965)

81 (Falconer D S 1967)

82 R A Fisher had done much of the groundwork in 1918(Fisher R A 1918).

83 (Denniston C 1998), (ICRP 2000)

84 (ICRP 2000)

85 The cancer risks are based on assessments by UNSCEAR, BEIR and ICRP and uses DDREF of 2 to convert risks from acute exposures to low does rate ones. See Annex B of ICRP 60 and p53 of ICRP103 for references. The

hereditary risk estimates come from the same sources. There are many cautions because the various numbers apply to rather different circumstances (and the differences are not always explicit) and include various weighting factors.

86 Based on Table 1 on page 53 (ICRP 2007)for the whole population.

## *10 A BETTER SPECIES?*

1 (Muller H J 1963) p256

2 (Muller H J 1933)

3 (Paul D 1988)

4 See (Wells H B 2012) p123

5 (deJong-Lambert, W 2014)

6 (Wells H B 2012)

7 (Anon 1953)

8 See (Galton, D J 1998) for the Greek contribution to eugenics

9 (Roper A G 1992)

10 J B S Haldane in *Man and his Future* p351-2 of(Wolstenholme G 1963)

11 Passage from Bible Communism (Oneida Community 1853) quoted by (Bajema CJ 1976) p49. He had written something similar in 1849 see (Noyes H H and Noyes G W 1923) and Bajema p78

12 Stirp: "The stock of a family; a line of descent; the descendants of a common ancestor" – a usage first recorded in 1502. *The Shorter Oxford English Dictionary*, 1973

13 (Noyes J H 1870)

14 (Noyes H H and Noyes G W 1923)

15 (McGee A N 1891)

16 (Galton F 1865a)

17 (Galton F 1865b)

18 (Galton F 1869)

19 (Galton F 1883)

20 (Galton F 1901)

21 The concluding words of his 1901 Huxley Lecture (Galton F 1901).

22 (Anon 1908)

23 (Galton F 1904)

24 These were people of of influence speaking against him. For example Maudsley was the distinguished psychiatrist commemorated by the hospital in his name; Benjamin Kidd an eminent sociologist, and Leslie Mackenzie, the Chief Medical Officer of Scotland.

25 (Horton T 2009)

26 (Webb S 1910)

# ENDNOTES

27 (Searle G R 1976) p107

28 McKenna had been responsible for the Dreadnought programme while at the Admiralty.

29 Mainly from (Searle G R 1976)

30 As reported by Charles Brace in the North American Review (Carlson E A 2001) p164

31 Jukes was a pseudonym adopted for the family by Dugdale. It was derived from a slang term for chickens which kept no permanent nests but laid their eggs promiscuously wherever fancy and convenience suggested.

32 (Goddard H H 1912) The Jackson Whites of New Jersey unaccountably escaped the attention of the sociologists and eugenicists. A mixed race community; they now prefer to be known as the Ramampough Mountain Indians.

33 Hypothyroidism is apparent in some mountainous regions because of the low levels of iodine in soil and water – a result of the element being washed out by snow and glacier melt. Most people's thyroid glands are active enough to concentrate sufficient iodine for normal bodily functions even under these circumstances. However, a few people have thyroids that are less active and these people suffer from goiters and the other problems when environmental iodine levels are low. Women with hypothyroidism tend to have children with hypothyroidism. Sufferers were know in the past as "cretins" but this term is now, like "lunatics", considered derogatory and is seldom used.

34 (Burbank L 1907) p82

35 (Webber H J and others 1914)

36 It was demolished in 1913.

37 p33 The square brackets indicate where there is some uncertainty in the transcription of the passage from Muller's notes.

38 (Carlson E A 2001)

39 Priddy had reportedly begun sterilising patients before the Act became law. In 1916 and 1917 he sterilised a total of 80 women – not all, it seems, feeble-minded. See http://www.uvm.edu/~lkaelber/eugenics/VA/VA.html. Virginia's Racial Integrity Act was passed on the same day as the Sterilization Act. This forbad white people from marrying other than white people.

40 (http://www.encyclopediavirginia.org/Petition to Sterilize Carrie Buck September 10 1924)

41 http://caselaw.lp.findlaw.com/scripts/getcase.pl?court=US&vol=274&invol=200 and http://readingroom.law.gsu.edu/cgi/viewcontent.cgi?article=1030&context=buckvbell

42 (Stern A M 2005)

43 The medal shows a standing man and woman in diaphanous clothing reaching towards a small naked infant – presumably hoping to evoke some kind of Greek ideal.

# GENES, FLIES, BOMBS...

44 (Rosen C 2004)

45 (Popenoe P 1934)

46 GHC Deb 21 July 1931 vol 255 cc1249-57 1249

47 Fitzgerald's own views are still debated.

48 (Kevles D J 1985) p69

49 (Muller H J 1933)

50 (Crow J F 1982)

51 (Muller H J 1935)

52 Elsewhere Lenin and Engels get a favourable mention.

53 Billy Sunday was a baseball player who became a popular evangelist in the early part of the 20$^{th}$ century. His revivalist meetings attracted large crowds until the radio established itself. Valentino was a movie heart-throb, Sullivan (the Boston Strong Boy) straddled the bare-knuckle and gloved eras of boxing in the final decades of the nineteenth century and became champion of both. Long was a controversial, ruthless and (to some) progressive politician who was assassinated in 1935 shortly after he declared that he would run for US president. Everyone knows who Al Capone was.

54 (Brewer H 1935)

55 (Crew F A E et al 1939) The Congress was adjourned just three days before war was declared.

56 This reflects the work of Sheila Weiss (Weiss S F n.d.)

57 (Mason J B 1938)

58 (Popenoe P 1934)

59 (Muller H J 1963) Muller was unable to attend because of illness.

## 11 A CHEMICAL GENETICS

1 Refer to Deichmann (Deichmann U 2007a) and (Deichmann U 2007b) and (Olby R C 1994). There is a good summary in (Authier A 2015) around p238

2 (Schwartz J 2008) p345 GRSp88 and, according to Dunn p182, it had also been proposed by Goldschmidt.

3 (Horowitz N H n.d.)

4 (Beadle G W 1939)

5 (Schrodinger E 1944). See (Dronamraju K R 1999) for a summary and review.

6 (Perutz M F 1987)

7 (Muller H J 1946)

8 (Stadler D 1997) David Ross Stadler, the author of the Perspectives review, is the son of Lewis Stadler and was for many years a Professor of Genetics at Washington State University. Other members of the family continued to have links with genetics.

# ENDNOTES

9 From the footnote on p23 of the Pilgrim Trust lecture (Muller H J 1947) written in early 1946.

10 Hershey and Chase 1952 (Hershey A D and Chase M 1952) For a summary of Griffith's and Hershey's work see (O'Connor C 2008).

11 (Chargaff E, Zamenhof S and Greene C 1950)

12 The ideas were put forward to a conference in Cambridge in September 1957 and published in 1958 (Crick F H C 1958). Crick published further thoughts on the Dogma in 1970.(Crick F 1970)

13 They were awarded the Nobel Prize in 1968 for their separate contributions to the problem.

## *12POSTSCRIPT*

1 Most of the information in these paragraphs is from (Carlson E A 1981).

# 14 GLOSSARY

*Allele*

One or more alternative forms of a particular gene.

*Autosome*

A chromosome that is not a sex chromosome. People normally have 22 pairs of autosomes (44 autosomes) in each cell together with two sex chromosomes (X and Y in the male and XX in the female).

*Centromere*

The point where the *chromatids* are joined after replication. The spindle – which draws the two chromatids apart at cell division – attaches to the centromere.

*Chromatids*

The two identical halves of a chromosome after replication in cell division that are joined at the *centromere*.

*Chromatin*

The combination of DNA and supporting proteins (histones) that make up the chromosomes.

*Chromosomes*

Chromosomes are a combination of DNA and small proteins called histones. The DNA is kept organised by being wound around the histones and then packed together. The chromosomes become visible at parts of the cell cycle.

Human cells contain 23 pairs of chromosomes: 22 pairs of autosomes and two sex chromosomes named X and Y. If the cell has two X chromosomes the person is female; if an X and a Y the person is male.

*Congenital*

Congenital disorders involve defects in or damage to a developing fetus. It may be the result of genetic abnormalities, the intra-uterine environment, errors of morphogenesis, or a chromosomal abnormality.

*Crossing over*

At an early stage of meiosis, after the chromosomes replicate,

# GLOSSARY

homologous pairs come together and randomly exchange segments. The resulting chromosomes are thus no longer either from the mother or father of the cell line but each one has elements of both. When the gametes are created, each one then has some genetic information from each of its parents in a form of *recombination*. Recombination and crossing over thus shuffles the pack of available genes and introduces variation into the genome.

### *Detriment*

Radiation detriment is defined by ICRP as: The total harm to health experienced by an exposed group and its descendants as a result of the group's exposure to a radiation source. Detriment is a multidimensional concept. Its principal components are the stochastic quantities:probability of attributable fatal cancer, weighted probability of attributable non-fatal cancer, weighted probability of severe heritable effects, and length of life lost if the harm occurs.

### *Dominant*

Most genes are present in two copies, one on each of the homologous chromosomes, one from the mother and one from the father. These may be different versions (alleles) of the gene. If a gene is expressed only when the same allele is present in both copies then it is said to be recessive. If an allele is expressed even if present only as a single copy then it is said to be dominant.

### *Effective dose*

Organs and tissues of the body have different sensitivities to radiation and each has been assigned a tissue weighting factor to account for that. For non-uniform irradiation (as when radioactive material gets into the body) the equivalent dose in each tissue or organ is multiplied by its weighting factor and this is then summed over all organs and tissues to give the effective dose. This then gives a measure of the total risk of the irradiation. Until quite recently the effective dose was known as the "effective dose equivalent". It is measured in Sieverts.

### *Epigenesis*

The development of a organism from a seed, spore or egg through a sequence of steps in which cells differentiate and form organs and tissue.

also

The theory that plants, animals and fungi develop in this way, in contrast to theories of preformationism.

# GENES, FLIES, BOMBS...

### Epigenetics

The study of the mechanisms of temporal and spatial control of gene activity during the development of complex organisms. Thus epigenetic can be used to describe any aspect other than DNA sequence that influences the development of an organism.

### Epilation

Hair loss caused by some chemical or physical agent

### Epistasis

The interaction between two or more genes.

### Equivalent dose

The result of multiplying absorbed dose dose by a weighting factor that represents the actual harm the type of radiation causes. For gamma and x-rays the weighting factor is unity, for neutrons about 10 and for alpha-particles 20.

### Eugenics

The improvement of the race by encouraging people considered to be of good quality to breed more and/or by encouraging, requiring or forcing those of lower quality to breed less or not at all.

### Eukaryote

Eukaryotes are organisms which have a distinct nucleus and nuclear membrane with DNA organised as chromosomes. All animals and plants are eukaryotes. Bacteria and archaea, which do not have a distinct nucleus and where DNA is a single strand, are known as prokaryotes.

### Eutelegenesis

Making available (by artificial insemination) the sperm of supposedly great men to suitable women in the expectation of improving the quality of the race.

### Gamete

A sex cell – sperm or egg or unpollinated seed.

### Genotype

The genetic constitution of an individual.

### Gray

The unit of absorbed dose – the energy absorbed per unit mass of tissue.

# GLOSSARY

One Gy= 1 J/kg.

### Heterosis

The increase in vigour resulting from hybridisation of pure-bred lines.

### Heterozygote

An individual who has inherited two different alleles of a particular gene. They are heterozygous for that gene and for the trait it is associated with.

### Homologous chromosome

Each eukarotic cell contains chromosomes in homologous pairs. Each chromosome of a homologous pair should have the same function and the same genes but, as one derives from the mother and the other from the father, the genes may be present as different alleles.

### Homozygote

An individual who has identical copies of the same gene on its two homologous chromosomes.

### Hybrids

The result of mating two distinct genetic stocks. These may be distinct species. So a mule is the offspring of a male donkey and a female horse and is, like many interspecies hybrids, virtually sterile. Mules and hinnies (a hinny is the the result of mating a male horse with a female donkey) have 63 chromosomes while horses have 64 and donkeys 62. Hybrids are also known between genera ( sheep and goats for example). Intraspecies hybrids are created when two subspecies are crossed (or varieties or cultivars in plants) and these are generally more likely to be fertile than interspecies crosses.

In plant breeding F1 (first filial) hybrids are the result of crosses between pure lines. A pure line is the result of breeding in which the best plant with the desired characteristics is self-pollinated, its seeds sown and grown on. The selection process is repeated until the plants are identical – the plants are homozygous. When two pure lines are crossed, the result is the F1 hybrid which may combine the desired characteristics of its parents. It may well be more vigorous (see heterosis) and it will be uniform. The F1 generation will not breed true so it is always necessary to bred from the pure lines – the breeder can thus, by holding onto these, recover his investment.

Almost all the corn (maize) and sugar beet grown is an F1 hybrid.

# GENES, FLIES, BOMBS...

### Inversion

The reversing of a section of DNA within the chromosome.

### Linked genes

Linked genes are those occurring on the same chromosome and hence being inherited together if it were not for *crossing over*.

### Mitosis

The splitting of a cell into two identical cells.

### Meiosis

Each normal cell contains two copies of each chromosome – one from the mother and one from the father. For sexual reproduction special cells, gametes, have to be produced with just one copy of each chromosome in a process called meiosis. Then when the two gametes (the sperm and the egg) fuse, the resulting cell has, once more, two copies of each chromosome, one from each of its parents. Meiosis takes place in two divisions and for each original cell four gametes are produced. As a result of *crossing over* each of these has a different genetic makeup.

### Neoplasm

An abnormal mass of tissue that results when cells divide more than they should or do not die when they should. Neoplasms may be benign (not cancer), or malignant (cancer). They are also called tumours

### Oocyte

An immature egg cell which, after meiosis, becomes an egg.

### Oophorectomy

The removal of the ovaries.

### Overdominance

A condition in genetics where the phenotype of the heterozygote lies outside the phenotypical range of both homozygous parents and may have a higher fitness than homozygous individuals.

### Phenotype

The observed phsyical form and behaviour of an organism, the result of its genotype and its interaction with its envirinment.

### Polymorphism

The existence within a population of two or more genotypes, the rarest

# GLOSSARY

of which exceeds some arbitrarily low frequency (say, 1 percent)

### Recombination

The exchange of DNA by crossing over in meiosis in eukaryotes, directly in bacteria and archaea and during some other cell processes such as DNA repair.

### Recessive

Most genes are present in two copies, one on each of the homologous chromosomes, one from the mother and one from the father. These may be different versions (alleles) of the gene. If a gene is expressed only when the same allele is present in both copies then it is said to be recessive. If an allele is expressed even if present as a single copy then it is said to be dominant.

### Salpingectomy

The removal of a Fallopian tube.

### Segregation

The separation of the homologous chromosomes during meiosis after crossing over and the sharing of genes between daughter cells.

### Sex chromosomes

Chromosomes that determine sex. In mammals denoted X and Y.

### Sievert

The unit of equivalent dose (and dose equivalent).

### Spermatogonia

Immature cells that turn into spermatozoa.

### Stochastic

Determined by chance. Stochastic effects of radiation which which may occur at any level of dose but their probability (and not their severity) is related to the dose. These include cancer and hereditary effects. In contrast deterministic effects occur whenever some threshold level is exceeded and their severity depends on the level of dose absorbed. The acute effects of radiation fall into this category as does the generation of cataracts.

## GENES, FLIES, BOMBS...

### *Translocation*

A section of DNA that is moved to another place in the chromosomes, generally to a non-homologous chromosome.

### *Wild type*

The form of a gene usually found in nature or that of a lab stock of normal individuals.

### *Zygote*

The cell formed when a sperm fertilises an egg. The cell contains genetic information from both parents and can divide and form an embryo.

# 15  BIBLIOGRAPHY

Adams M, 1968. The founding of population genetics:contributions of the Chetverikov school, 1924-1934. *J Hist Biol*, 1(1), pp.23–39.

Alexander M L, 1954. Mutation rates at specific autosomal loci in the mature and immature germ cells of Drosophila Melanogaster. *Genetics*, 39, pp.409–427.

Altenburg E and Muller H J, 1920. The genetic basis of truncate wing - an inconstant and modifiable character in Drosophila. *Genetics*, 5, pp.1–59.

Anon, 1947. Genetic Effects of the Atomic Bombs in Hiroshima and Nagasaki (Genetics Conference, Comm on Atomic Casualties, NRC, June 1947). *Science*, 106(2754), pp.331–333.

Anon, 1953. Muller Opposes Reds in Schools. *The Daily Banner, 16 March*.

Anon, 1908. Report of the Royal Commission on the care and control of the feeble-minded. *The Times, 1 August*, p.5.

Anon, 1960. *The Biological Effects of Atomic Radiation. Summary Report from a Study by the National Academy of Sciences.*, National Academy of Sciences USA.

Anon (Chambers R), 1844. *Vestiges of the Natural History of Creation*, London: John Churchill. Available at: http://www.esp.org/books/chambers/vestiges/facsimile/.

Auerbach C, 1976. *Mutation Research: Problems, results and perspectives*, London: Chapman & Hall. Available at: https://books.google.co.uk/books?id=jlf0BwAAQBAJ&pg=PA3&dq=auerbach+presence+absence&hl=en&sa=X&ved=0CC4Q6AEwA2oVChMI--G1yvD9xwIVC9UaCh2TfAss#v=onepage&q=auerbach%20presence%20absence&f=false.

Authier A, 2015. *Early Days of X-ray Crystallography*, Oxford: OUP. Available at:

https://books.google.co.uk/books?
id=NC6_4yip49MC&pg=PA238&lpg=PA238&dq=bragg+anthracene&so
urce=bl&ots=WUXsOdXMSU&sig=bsRHOgTE6BFov6yY2CFP-
231AWc&hl=en&sa=X&redir_esc=y#v=onepage&q=bragg
%20anthracene&f=false.

Awa A A, 1975. Review of thirty years study of Hiroshima and Nagasaki atomic bomb survivors. II Biological effect. B Genetic effects. 2 Cytogenetic study. *J Radiat Res Suppl*, 16, pp.75–81.

Awa A A, Honda T, Sofuni T, Nerrishi S, Yoshida M C and Matsui T, 1971. Chromosome-aberration frequency in cultured blood-cells in relation to radiation dose of a-bomb survivors. *Lancet*, 298, pp.903–905.

Bajema CJ ed., 1976. *Eugenics: Then and Now*, Stroudsberg, PA: Dowden, Hutchinson and Rossi.

Balzer F, 1967. *Theodor Boveri: The Life of a Great Biologist 1862-1915*, Berkeley: U Cal Press. Available at: http://zygote.swarthmore.edu/fert6b.html [Accessed December 26, 2009].

Bardeen C R, 1907. Abnormal development of toad ova fertilized by spermatozoa exposed to the Roentgen rays. *J Exptl Zool*, 4(1), pp.1–44.

Barker M and Lerner I M, 1961. Translation of: On certain aspects of the evolutionary process from the standpoint of modern genetics, Chetverikov 1926 paper by M. Barker and I. M. Lerner. *Proc Am Phil Soc*, 105, pp.167–195.

Bateson W, 1909. *Mendel's Principles of Heredity*, Cambridge: CUP.

Bateson W, Saunders E R and Punnett R C, 1904. Reports to the Evolution Committee of the Royal Society. Report II: Experimental studies in the physiology of heredity.

Beadle G W, 1939. Physiological aspects of genetics. *Ann Rev Physiology*, 1, pp.41–62.

BEAR Committee, 1956a. Genetic effects of atomic radiation. *Science N S*, 123(3209), pp.1157–1164.

# BIBLIOGRAPHY

BEAR Committee, 1956b. *The Biological Effects of Atomic Radiation*, Washington DC: National Academy of Sciences/NRC. Available at: http://hdl.handle.net/2027/mdp.39015049805065.

Beebe G W, 1979. Reflections on the work of the ABCC in Japan. *Epidemiol Revs*, 1, pp.184–210.

Beebe G W, Ishida M, Jablon S, 1962. Studies of the mortality of A-bomb survivors. 1. Plan of study and mortality in the medical subsample (Selection 1), 1950-1958. *Radiat Res*, 16, pp.253–280.

BEIR Committee, 1990. *Health effects of exposure to low levels of ionizing radiation: BEIR V*, Washington DC: National Academy Press / No:

BEIR Committee, 2006. *Health Risks from Exposure to Low Levels of Ionizing Radiation BEIR VII Ph2*, National Research Council / No:

BEIR Committee, 1972. *The effects on populations of exposure to low levels of ionizing radiation*, National Research Council / No:

BEIR Committee, 1979. *The effects on populations of exposure to low levels of ionizing radiation*, National Research Council / No:

BEIR Committee, 1980. *The effects on populations of exposure to low levels of ionizing radiation*, National Research Council / No:

Bennett M & Grunfeld A T (ed), 2004. *On Her Own: Journalistic Adventures From San Francisco to the Chinese Revolution 1917-1927*, Beijing: Foreign Language Press.

Benson K R, 1991. Observation versus theory: research in transition. In Dinsmore C E, ed. *A History of Regeneration Research*. Cambridge: CUP, pp. 91–100.

Bergonie J and Tribondeau L, 1906. De quelques resultats de radiotherapie et essai de fixation d'une technique rationnelle. *C R Acad Sci(Paris)*, pp.985, 983.

Binder O O, 1953. How nuclear radiation can change our race. *Mechanix Illustrated*, (December), p.108-.

# GENES, FLIES, BOMBS...

Bohr N, 1933. Light and life. *Nature*, 131, p.(3308)421-423 and (3309)457-459.

Bourguet D, 1999. The evolution of dominance. *Heredity*, 83, pp.1–4.

Boveri T, 1929. *The Origin of Malignant Tumors*, Baltimore,MD: Williams and Wilkins.

Boveri T, 1914. *Zur Frage der Entstehung maligner Tumoren*, Jena, Germany: Gustav Fischer Verlag.

Brewer H, 1935. Eutelegenesis. *Eugenics Rev*, 27(2), pp.121–126.

Bridges C B, 1936. The bar "gene", a duplication. *Science*, 83, pp.210–211.

Browne J, 1989. Botany for gentlemen: Erasmus Darwin and "The Loves of the Plants." *Isis*, 80(4), pp.593–621.

Burbank L, 1907. *The Training of the Human Plant*, New York: The Century Co.

Cantril S T and Parker H M, 1945. *The tolerance dose*, US AEC / No: MDDC-100.

Carlson E A, 1981. *Genes, Radiation and Society: The Life and Work of H J Muller*, Ithaca;London: Cornell UP.

Carlson E A, 2011. Speaking Out About the Social Implications of Science: The Uneven Legacy of H. J. Muller. *Genetics*, 187, pp.1–7.

Carlson E A, 2001. *The Unfit: A History of a Bad Idea*, Cold Spring Harbor, NY: Cold Spring Harbor Lab Press.

Carter T C, 1959. A pilot experiment with mice, using Haldane's method for detecting induced autosomal recessive lethal genes. *J Genet*, 56, pp.353–362.

Carter T C, Lyon M F and Phillips R J S, 1956. Induction of mutations in mice by chronic gamma irradiation: interim report. *Radiology*, 29, p.106-.

Castle W E, 1919a. Are genes linear or non-linear in arrangement? *Proc N A S*, 5(11), pp.500–506.

Castle W E, 1919b. Is the arrangement of genes in the chromosome linear? *Proc N*

*A S*, 5(2), pp.25–32.

Chakraborty R N and Yasuda C, 1998. Ionizing radiation and genetic risk VII The concept of mutation component and its use in risk estimation for Mendelian diseases. *Mutat Res*, 400, pp.541–552.

Chargaff E, Zamenhof S and Greene C, 1950. Composition of human desoxypentose nucleic acid. *Nature*, 165(4202), pp.756–757.

Charles D R, 1950. Radiation-induced mutations in mammals. *Radiology*, 55, pp.579–581.

Charles D R, Tihen J A, Otis E M and Grobman A B, 1961. Genetic effects of chronic x-irradiation in mice. *Genetics*, 46, pp.5–8.

Clarke G N, 2006. A R T in history, 1678-1978. *Hum Reprod*, 21(7), pp.1645–1650.

Cobb M, 2002. Malpighi, Swammerdam and the Colourful Silkworm: Replication and Visual Representation in Early Modern Science. *Ann Science*, 59, pp.111–147.

Cobb M, 2000. Reading and writing The Book of Nature: Jan Swammerdam(1637-1680). *Endeavour*, 24(3), pp.122–128.

Cock A G, 1983. William Bateson's rejection and eventual acceptance of chromosome theory. *Ann Science*, 40, pp.19–59.

Cole F J, 1930. *Early Theories of Sexual Generation*, Oxford: OUP.

Colwell H A and Russ S, 1934. *X-ray and radium injuries. Prevention and treatment*, London: OUP.

Coppes-Zantinga A R and Coppes M J, 1998. The early years of radiation protection: a tribute to Madame Curie. *Can Med Assoc J*, 159, pp.1389–1391.

Court Brown W M and Doll R, 1957. London:HMSO / No: MRC Special Report 295.

Crew F A E, 1969. Recollections of the early days of the Genetical Society. In Jinks J, ed. *The Genetical Society - The First Fifty Years*. Edinburgh: Oliver and Boyd, pp. 9–15.

Crew F A E and Lamy R, 1935. *The Genetics of the Budgerigar*, Bradford and London: Watmoughs.

Crew F A E et al, 1939. The geneticists' manifesto. *J Hered*, 30, pp.371–373.

Crick F, 1970. Central dogma of molecular biology. *Nature*, 227, pp.561–563.

Crick F H C, 1958. On protein synthesis. *Symp Soc Exptl Biol*, 12, pp.138–163.

Crow E W and Crow J F, 2002. 100 tears ago: Walter Sutton and the chromosome theory of heredity. *Genetics*, 160, pp.1–4.

Crow J F, 1998. 90 years ago: The beginning of hybrid maize. *Genetics*, 148, pp.923–928.

Crow J F, 2006. H J Muller and the "Competition Hoax." *Genetics*, 173, pp.511–514.

Crow J F, 2008. Mid-century controversies in population genetics. *Annu Rev Genet*, 42, pp.1–16.

Crow J F, 1995. Quarreling geneticists and a diplomat. *Genetics*, 140, pp.421–426.

Crow J F, 1982. Review of. Genes, radiation and society. *Am J Hum Genet*, 34, pp.519–523.

Crow J F, 1992. Sixty years ago: the 1932 International Congress of Genetics. *Genetics*, 131, pp.761–768.

Darden L, 1991. *Theory Change in Science: Strategies from Mendelian Genetics*, OUP. Available at: http://books.google.co.uk/books?id=O_-jPuVdoo8C&pg=PA42&lpg=PA42&dq=origins+mendelism&source=bl&ots=NzARVE2U69&sig=PAOnPQELh4EK6Z7m2V3Xoty4nw4&hl=en&ei=JAXwSq-QNsefjAfx2N3ICA&sa=X&oi=book_result&ct=result&resnum=9&ved=0CCoQ6AEwCDgU#v=onepage&q=origins%20mendelism&f=false.

Darwin C R, 1868. *The Variation of Animals and Pants under Domestication* 1st ed., London: John Murray. Available at: http://darwin-

# BIBLIOGRAPHY

online.org.uk/content/frameset?
itemID=F877.2&viewtype=text&pageseq=1.

Davis A P and Justice M J, 1998. An Oak Ridge Legacy: The Specific Locus Test and Its Role in Mouse Mutagenesis. *Genetics*, 148, pp.7–12.

de Chadarevian, S, 2006. Mice and the Reactor: The "Genetics Experiment" in 1950s Britain. *J Hist Biol*, 39, pp.707–735.

Deichmann U, 2007a. A brief review of the early history of genetics and its relationship to physics and chemistry. In *Max Delbrück and Cologne. An Early Chapter of German Molecular Biology*. NJ, USA: World Scientific, pp. 3–18. Available at: http://www.worldscientific.com/doi/pdf/10.1142/9789812775818_fmatter.

Deichmann U, 2007b. "Molecular" versus "Colloidal": Controversies in biology and biochemistry, 1900–1940. *Bull Hist Chem*, 32(2), pp.105–118.

deJong-Lambert, W, 2014. *The Cold War Politics of Genetics Research*, Dordrecht, NL: Springer.

Denniston C, 1982. Low level radiation and genetic risk estimation in man. *Ann Rev Genetics*, pp.329–355.

Denniston C, Chakraborty R and Sankaranarayanan K, 1998. Ionizing radiation and genetic risk VII The concept of mutation component and its use in risk estimation for multifactorial diseases. *Mutat Res*, 405, pp.57–79.

Dobzhansky T, 1937. *Genetics and the Origin of Species*, Columbia University Press.

Doll R, 1995. Hazards of ionizing radiation - 100 years of observation on man. *Brit J Cancer*, 72, pp.1339–1349.

Dronamraju K R, 1999. Erwin Schrodinger and the Origins of Molecular Biology. *Genetics*, 153, pp.1071–1076.

Dunn L C, 1965. *William Ernest Castle: A biographical memoir*, US National Academy of Sciences.

# GENES, FLIES, BOMBS...

East E M, 1936. Heterosis. *Genetics*, 21, pp.375–397.

Edwards A W F, 2008. G H Hardy (1908) and Hardy-Weinberg equilibrium. *Genetics*, pp.1143–1150.

Eisenbud M and Harley J T, 1958. Long-term fallout. *Science N S*, 128(3321), pp.399–402.

Evans J A, 1998. *Celluloid Mushroom Clouds: Hollywood and the Atomic Bomb*, Boulder CO;Oxford: Westview Press.

Falconer D, 1993. Quantitative Genetics in Edinburgh: 1947-1 980. *Genetics*, 133, pp.137–142.

Falconer D S, 1965. The inheritance of liability to certain diseases estimated from the incidence among relatives. *Ann Hum Genet*, 29, pp.51–76.

Falconer D S, 1967. The inheritance of liability to diseases with variable age of onset with particular reference to diabetes mellitus. *Ann Hum Genet*, 31, pp.1–20.

Felsenstein J, 1974. The evolutionary advantage of recombination. *Genetics*, 78, pp.737–756.

Fisher R A, 1918. The correlation between relatives on the supposition of Mendelian inheritance. *Trans Roy Soc Edinburgh*, pp.433, 399.

Fisher R A, 1930. *The Genetical Theory of Natural Selection*, Oxford: Clarenden Press.

Folley J H, Borges W and Yamawaki T, 1952. Incidence of leukemia in survivors of the atomic bomb in Hiroshima and Nagasaki,Japan. *Am J Med*, 13(3), pp.311–321.

Galton, D J, 1998. Greek theories on eugenics. *J Med Ethics*, 24, pp.263–267.

Galton F, 1904. Eugenics: Its definition, scope, and aims. *Am J Sociol*, 10(1).

Galton F, 1869. *Hereditary Genius - An Inquiry into its Laws and Consequences* 1st ed., London: MacMillan.

# BIBLIOGRAPHY

Galton F, 1865a. Hereditary talent and character PtI. *Macmillan's Mag*, 12(68), pp.157–1966.

Galton F, 1865b. Hereditary talent and character PtII. *Macmillan's Mag*, 12(71), pp.318–327.

Galton F, 1883. *Inquiries into Human Faculty and its Development*, MacMillan.

Galton F, 1901. The possible improvement in the human breed. *Lecture Anthropological Institute, London, 29 October*.

Gillham N W, 2001. Evolution by Jumps: Francis Galton and William Bateson and the Mechanism of Evolutionary Change. *Genetics*, 159, pp.1383–1392.

Glad J, 2003. Hermann J Muller's 1936 letter to Stalin. *Mankind Quart*, 43(3), pp.305–319.

Gliboff S, 1999. Gregor Mendel and the laws of evolution. *Hist Sci*, 37, pp.217–235.

Goddard H H, 1912. *The Kallikak Family: A Study of the Heredity of Feeble-Mindedness*, London; NY: MacMillan Publishing Co.

Goldschmidt R B, 1946. Position effect and the theory of the corpuscular gene. *Experientia*, 2, pp.197–232.

Graham L R, 1972. *Science and Philosophy in the Soviet Union* 1st ed., NY: Alfred A Knopf.

Graham L R, 1993. *Science in Russia and the Soviet Union: a short history*, Cambridge;NY: Cambridge University Press.

Graham L R, 1987. *Science, Philosophy and Human Behavior in the Soviet Union* 2nd ed., Columbia University Press.

Grant B S, 1999. Fine tuning the Peppered Moth paradigm. *Evolution*, 53, pp.980–984.

Green J and LaDuke J, 2009. *Pioneering women in American mathematics: the pre-1940 PhDs*, Am Math Soc.

Grobman A B, 1952. Letter to the Editor. *Am J Human Genet*, 4(1), pp.54–55.

# GENES, FLIES, BOMBS...

Grobman A B, 1951. *Our Atomic Heritage*, Gainsville: Univ Florida Press.

Haldane J B S, 1924. A mathematical theory of natural and artificial selection Part 1. *Trans Camb Phil Soc*, 23, pp.19–41.

Haldane J B S, 1955. Genetical Effects of Radiation from Products of Nuclear Explosions. *Nature*, 176, p.115.

Haldane J B S, 1956. The detection of autosomal lethals in mice induced by mutagenic agents. *J Genet*, 54, pp.56–63.

Hamblin J D, 2007. A Dispassionate and Objective Effort:' Negotiating the First Study on the Biological Effects of Atomic Radiation. *J Hist Biol*, 40(1), pp.147–177.

Hanson F B and Heys F, 1929. An analysis of the effects of the different rays of radium in producing lethal mutations in Drosophila. *Amer Nat*, 63, pp.201–213.

Harada T and Ishida M, 1960. Neoplasms Among A-Bomb Survivors in Hiroshima: First Report of the Research Committee on Tumor Statistics, Hiroshima City Medical Association, Hiroshima, Japan. *J Natl Cancer Inst*, 25, pp.1253–1264.

Harris H, 1998. *The Birth of the Cell*, New Haven: Yale Univ Press.

Hershey A D and Chase M, 1952. Independent functions of viral protein and nucleic acid in growth of bacteriophage. *J Gen Physiol*, 36(1), pp.39–56.

Hilbert D, 1930. Naturkennen und Logik. *Naturwissenschaften*, pp.959–963.

Horowitz N H, Neurospora and the beginnings of molecular genetics. Available at: https://web.stanford.edu/group/neurospora/UsefulPDFs/20Horowitz.pdf.

Horton T, 2009. A short guide to the Minority Report. In *From the Workhouse to Welfare*. Fabian Society, pp. 9–20.

ICRP, 1951. International recommendations on radiological protection. *Br J Radiol*, 23, pp.46–53.

# BIBLIOGRAPHY

ICRP, 2000. *Publication 83 Risk estimation for multifactorial diseases*, ICRP.

ICRP, 1955. Recommendations of the International Commission on Radiological Protection. *Br J Radiol*, p.o.

ICRP, 1964. *Recommendations of the International Commission on Radiological Protection Publication 6*, London: Pergammon.

ICRP, 2007. *The 2007 recommendations of the International Commission on Radiological Protection: Publication 103*, ICRP.

IXRPC, 1934. International recommendations for X-ray and radium protection. *Br J Radiol*, 7, pp.695–699.

Jablon S and Kato H, 1972. Studies of the Mortality of A-Bomb Survivors: 5. Radiation Dose and Mortality, 1950-1970. *Radiat Res*, 50, pp.649–698.

Jablon S, Ishida M and Beebe G W, 1964. Studies of the mortality of A-bomb survivors 2.Mortality of selections I and II. *Radiat Res*, 21, pp.423–445.

Joint Comm Atomic Energy, 1957. *Summary - Analysis of Hearings May 27-29, and June 3-7, 1957 on the Nature of Radioactive Fallout and its Effects on Man*, Washington DC: US Govt Print. Off. Available at: Stanford.

Jones C G, 2005. A review of the history of U.S. radiation protection regulations, recommendations and standards. *Health Physics*, pp.124, 105.

Joravsky D, 1970. *The Lysenko Affair*, Cambridge, MA: Harvard UP.

Kato H and Schull W J, 1982. Studies of the mortality of A-bomb survivors Report 7 Mortality 1950-1978 Part1 Cancer mortality. *Radiat Res*, 90(2), pp.395–432.

Kato H, Schull W J and Neel J V, 1966. A Cohort-Type Study of Survival in the Children of Parents Exposed to Atomic Bombings. *Am J Human Genet*, 18, pp.339–373.

Kevles D J, 1985. *In the Name of Eugenics*, Berkeley: Univ Cal Press.

Kirschenbaum L A, 2015. *International Communism and the Spanish Civil War*, NY: Cambridge University Press.

# GENES, FLIES, BOMBS...

Linzie A, 2006. *The True Story of Alice B. Toklas: A Study of Three Autobiographies*, Univ Iowa Press.

Lorenz E, 1950. Some biologic effects of long-continued irradiation. *Am J Roentgenology*, 63, p.176.

Lovejoy A O, 1936. *The Great Chain of Being: a study of the history of an idea*, Harvard UP. Available at: http://books.google.co.uk/books?id=5u3HZjTpkTgC&dq=chain+of+being&printsec=frontcover&source=bl&ots=jzulcmKCSB&sig=oPsEorWJl-ey-FhVqMLhu4F-i9o&hl=en&ei=R2LtSvWLMqHSjAeXycSeDQ&sa=X&oi=book_result&ct=result&resnum=3&ved=0CBgQ6AEwAg#v=onepage&q=&f=false.

Lyon M and Morris T, 1966. Mutation rates at a new set of specific loci in the mouse. *Genet Res*, 7, pp.12–17.

MacKenzie K and Muller H J, 1940. Mutation effects of ultra-violet light in Drosophila. *Proc Roy Soc B*, 129, pp.491–516.

Maderspacher F, 2008. Theodor Boveri and the natural experiment. *Current Biology*, 18(7), pp.R279–R286.

Magner L N, 1979. *A History of the Life Sciences*, NY: M Dekker.

March H C, 1944. Leukemia in radiologists. *Radiology*, 43(3), pp.275–278.

March H C, 1950. Leukemia in radiologists in a 20 year period. *Am J Med Sci*, 220, pp.282–286.

Martland H S, 1929. Occupational poisoning in manufacture of luminous watch dials. *J Am Med Assoc*, 92(6&7), pp.466–473&552–559.

Martland H S, 1931. The occurrences of malignancy in radioactive persons. *Am J Cancer*, 15(4), pp.2435–2516.

Martland H S, Conlon P and Knef J P, 1925. Some unrecognised dangers in the use and handling of radioactive substances with especial reference to the storage of insoluble products of radium and mesothorium in the reticula-endothelial system. *J Am Med Assoc*, 85(23), pp.1769–1775.

Mason J B, 1938. Germany Tries Sterilization. *Soc Science*, 13(4), pp.303–309.

# BIBLIOGRAPHY

Mavor J W, 1923. An effect of X-rays on the linkage of Mendelian characters in the first chromosome of Drosophila. *Genetics*, 8(4), pp.355–366.

Mavor J W, 1922. The production of non-disjunction by X-rays. *Science*, 55(1420), pp.295–297.

Mavor J W, 1924. The production of non-disjunction by x-rays. *J Exptl Zoology*, 39(2), pp.381–432.

Mayr E, 1982. *The Growth of Biological Thought*, Cambridge MA: Harvard UP.

Mayr E, 1985. Weismann and evolution. *J Hist Biol*, 18(3), pp.295–329.

Mayr E and Provine W B (eds) ed., 1980. *The Evolutionary Synthesis: perspectives on the unification of biology*, Cambridge MA;London: Harvard UP.

Mazzarello P, 1999. A unifying concept: the history of cell theory. *Nature Cell Biology*, 1(May), pp.E13–E15.

McGee A N, 1891. An experiment in human stirpiculture. *Am Anthropologist*, 4(4), pp.319–326.

McGregor J H, 1908. Abnormal development of frog embryos as a result of treatment of ova and sperm with Roentgen rays. *Science*, p.?, 445.

McKaughan D J, 2005. Revisiting the hopes inspired by "Light and Life." *Isis*, 96, pp.507–529.

Meggitt G C, 2008. *Taming the Rays: a history of radiation and protection*, Lulu.com.

Mendel G, 1866. Versuche über Plflanzenhybriden (1865)(trans:Experiments in plant hybridization. *Verhandlungen des naturforschenden Vereines in Brünn*, IV, pp.3–47.

Michurin I V, 1934. Results of My Sixty Years' Work and prospects for the Future. *Trans I V Michurin Plant Breeding Station*, 2. Available at: http://www.marxists.org/reference/archive/michurin/works/1930s/results.htm [Accessed November 17, 2010].

Mole R H, 1959. Some aspects of mammalian radiobiology. *Radiat Res Supp*, 1, pp.124–148.

Moll W, 2006. WAW Moll Antonie van Leeuwenhoek. *www.euronet.nl/users/warnar/leeuwenhoek.html*. Available at: http://www.euronet.nl/users/warnar/leeuwenhoek.html [Accessed October 18, 2009].

Moloney W C and Kastenbaum M A, 1955. Leukemogenic effects of ionizing radiation on atomic bomb survivors in Hiroshima city. *Science*, p.308,.

Morgan T H, 1910a. Chromosomes and heredity. *Amer Nat*, 44, pp.449–496.

Morgan T H, 1910b. The method of inheritance of two sex-limited characters in the same animal. *Proc Soc Exper Biol Med*, 8, pp.17–19.

Morgan T H, 1934. The relation of genetics to physiology and medicine - Nobel Lecture.

Morgan T H, Sturtevant A H, Muller H J and Bridges C B, 1915. *The Mechanism of Mendelian Inheritance*, London: Constable.

Mould R F, 1995. The early history of x-ray diagnosis with emphasis on the contributions of physics 1895-1915. *Phys Med Biol*, 40(11), pp.1741–1787.

MRC, 1956. *The hazards to man of nuclear and allied radiations*, London:HMSO / No: Cmnd 9780.

Muckerheide J, 2000. Applying Radiation Health Effects Data to Radiation Protection Policies. In *IRPA 10*. IRPA Congress. Hiroshima.

Muller H J, 1946. A physicist stands amazed at genetics. *J Hered*, 37, pp.90–92.

Muller H J, 1940. An analysis of the process of structural change in chromosomes of Drosophila. *J Genetics*, 40(1&2), p.1.

Muller H J, 1927. Artificial transmutation of the gene. *Science*, pp.84–87.

Muller H J, 1936. Basis of the theory of the gene: the experimental evidence concerning the properties of the gene. Summary of an address given at the Lenin Academy of Agricultural Science, 23 December 1936.

# BIBLIOGRAPHY

Muller H J, 1926. Determining identity of twins. *J Hered*, 17, pp.195–206.

Muller H J, 1943. Edmund B. Wilson -- an appreciation. *Am Nat*, 77, pp.5–37 &142–172.

Muller H J, 1963. Genetic Progress by Voluntarily Conducted Germinal Choice. In *Man and his Future*. Boston; Toronto: Little,Brown and Co.

Muller H J, 1949. Genetics in the scheme of things. *Hereditas*, 35(Supplement), pp.96–127.

Muller H J, 1958. How much is evolution accelerated by sexual reproduction? *Anat Rec*, 132, pp.480–481.

Muller H J, 1925. Mental traits and heredity as studied in a case of identical twins reared apart. *J Heredity*, 16, pp.433–448.

Muller H J, 1964a. Muller publication list from Studies in Genetics: "The selected Papers of H J Muller."

Muller H J, 1948a. Mutational prophylaxis. *Bull NY Acad Med*, 24, pp.447–469.

Muller H J, 1923. Observations of Biological Science in Russia. *Sci Monthly*, 16, pp.539–552.

Muller H J, 1950. Our load of mutations. *Am J Hum Genet*, 2(2), pp.111–176.

Muller H J, 1935. *Out of the Night: A biologist's view of the future*, New York: Vanguard Press.

Muller H J, 1939. *Report of Investigation with Radium*, London: Medical Research Council.

Muller H J, 1951. Science in bondage. *Science*, 113, pp.25–29.

Muller H J, 1932. Some genetic aspects of sex. *Am Naturalist*, 66(703), pp.118–138.

Muller H J, 1964b. *Studies in Genetics: The selected Papers of H J Muller*, Bloomington, IN: Indiana UP.

Muller H J, 1948b. The crushing of genetics in the USSR. *Bull Atomic Scientists*, 12(December), pp.369–371.

Muller H J, 1933. The dominance of economics over eugenics. *Scientific Monthly*, 37(1), pp.40–47.

Muller H J, 1947. The gene - Pilgrim Trust Lecture. *Proc Roy Soc B*, 134, pp.1–37.

Muller H J, 1928. The problem of genic modification. In *Proc 5th Int Congress of Genetics*. 5th Int Congress of Genetics, Sept 1927. Berlin, pp. 234–260.

Muller H J, 1964c. The relation of recombination to mutational advance. *Mutat Res*, 1, pp.2–9.

Muller H J, 1938. The remaking of chromosomes. *Collecting Net (Woods Hole)*, 13, pp.181–198.

Muller H J and Jacobs-Muller J M, 1925. The standard errors of chromosome distances and coincidence. *Genetics*, 10, pp.509–524.

Muller H J and MacKenzie K, 1939. Discriminatory effect of ultra-violet rays on mutation in Drosophila. *Nature*, 143, pp.83–84.

Neel J V, 1999. Changing Perspectives on the Genetic Doubling Dose of Ionizing Radiation for Humans, Mice, and Drosophila. *Teratology*, 59, pp.216–221.

Neel J V, 1951. Our Atomic Heritage (review of). *Am J Human Genet*, 3(1), pp.81–83.

Neel J V, Kato H and Schull W J, 1974. Mortality in the children of atomic bomb survivors and controls. *Genetics*, 76, pp.311–326.

Neel J V & Schull W J, 1956. *The effect of exposure to the atomic bombs on pregnancy terminationin Hiroshima and Nagasaki*, Hiroshima/Wash DC: ABCC/NAS. Available at: http://www.nap.edu/openbook.php?record_id=1800&page=13.

Noyes H H and Noyes G W, 1923. The Oneida Community experiment in stirpiculture. In Davenport C B et al, ed. *Eugenics, Genetics and the Family*. Baltimore: The Williams and Wilkins Co, pp. 374–386.

# BIBLIOGRAPHY

Noyes J H, 1870. Scientific propagation. In Goodman D, ed. *The Modern Thinker*. Am News Co, pp. 97–120.

O'Connor C, 2008. Isolating Hereditary Material: Frederick Griffith, Oswald Avery, Alfred Hershey, and Martha Chase. *Nature Education*, 1(1), p.105.

Olby R, 1985. *Origins of Mendelism* 2nd ed., Chicago: Univ Chicago Press.

Olby R C, 1997. Mendel, Mendelism and Genetics. Available at: http://www.mendelweb.org/MWolby.html#s3.

Olby R C, 1994. *The Path to the Double Helix* 2nd ed., Dover. Available at: https://books.google.co.uk/books?id=AGLDAgAAQBAJ&pg=PA114&lpg=PA114&dq=gene+klampenborg&source=bl&ots=jItpBc1RBm&sig=5b5_osrl4wCiRntRPUFcxpPCnmw&hl=en&sa=X&redir_esc=y#v=onepage&q=gene%20klampenborg&f=false.

Ozasa K , Shimizu Y , Suyama A , Kasagi F, Midori Soda M ,Grant E J , Sakata R, Sugiyama H and Kodama K, 2012. Studies of the Mortality of Atomic Bomb Survivors, Report 14, 1950–2003: An Overview of Cancer and Noncancer Diseases. *Radiat Res*, 177, pp.229–243.

Paigen K, 2003a. One Hundred Years of Mouse Genetics: An Intellectual History. I. The Classical Period (1902–1980). *Genetics*, 163, pp.1–7.

Paigen K, 2003b. One Hundred Years of Mouse Genetics: An Intellectual History. II. The Molecular Revolution (1981–2002). *Genetics*, 163, pp.1227–1235.

Palfreeman L, 2015. *Spain Bleeds: The Development of Battlefield Blood Transfusion During the Civil War*, Eastbourne, UK: Sussex Academic Press.

Paul D, 1988. H J Muller, communism and the cold war. *Genetics*, 119, pp.223–225.

Paul D B, 1983. A War on Two Fronts: J. B. S. Haldane and the Response to Lysenkoism in Britain Author(s): Diane B. Paul Source: Journal of the History of Biology, Vol. 16, No. 1 (Spring, 1983), pp. 1-37. *J Hist Biol*, 16(1), pp.1–37.

Perutz M F, 1987. Physics and the riddle of life. *Nature*, 326, pp.555–558.

Phillips P C, 1998. The language of gene interaction. *Genetics*, 149(3), pp.1167–1171.

Plummer G, 1952. Anomalies occurring in children exposed in utero to the atomic bomb at Hiroshima. *Pediatrics*, 10, pp.692–687.

Popenoe P, 1934. The German sterilization law. *J Heredity*, 25, pp.257–260.

Preston D L, Shimizu Y, Pierce D A Suyama A and Mabuchi K, 2003. Studies of mortality of atomic bomb survivors. Report 13: Solid cancer and noncancer disease mortality: 1950-1997. *Radiat Res*, 160, pp.381–407.

Preston D L, Shimizu Y, Pierce D A, Suyama A and Mabuchi K, 2003. Studies of Mortality of Atomic Bomb Survivors. Report 13: Solid Cancer and Noncancer Disease Mortality: 1950–1997. *Radiat Res*, 160, pp.381–407.

Preston P, 2012. *We Saw Spain Die*, Hachette UK.

Provine W B, 1980. Genetics. In Mayr E and Provine W B, ed. *The Evolutionary Synthesis*. Camb,MS;London: Harvard UP.

Provine W B, 1971. *The Origins of Theoretical Population Genetics*, Chicago: University Chicago Press.

Punnett R C ed., 1941. *Proceedings of the Seventh Intenational Genetical Congress Edinburgh Scotland 23 30 August 1939*, London: Cambridge University Press.

R C, 1941. Proceedings of the Edinburgh Genetics Congress. *J Heredity*, 32(12), pp.426–428.

Rader K A, 2004. *Making Mice: Standardizing Animals for American Biomedical Research 1900-1955*, Princeton Univ Press.

Ray-Chaudhuri S P, 1939. The validity of the Bunsen-Roscoe law in the production of mutations by radiation of extremely low intensity. In *Proceedings of the Seventh International Genetical Congress ed R C Punnett*. Seventh International Genetical Congress. Edinburgh: Cambridge University Press, p. 246.

Regaud C and Dubreuil G, 1908. Perturbations dans le developpement des oeufs fecondes par des spermatozoides rontgenises chez le lapin. *C R Soc Biol(Paris)*, 64, pp.1014–1016.

Rhoades M M, 1957. Lewis John Stadler 1896-1954. In *Biographical Memoirs*. National Academy of Sciences USA.

Robinson G, 1979. *A Prelude to Genetics*, Lawrence, KA, USA: Coronado Press Inc.

Roll-Hansen N, 2005. The Lysenko effect:undermining the autonomy of science. *Endeavour*, 29(4), pp.143–147.

Roper A G, 1992. Ancient eugenics. *Mankind Quart*, 32, pp.383–418.

Rosen C, 2004. *Preaching Eugenics: Religious Leaders and the American Eugenics Movement*, NY: OUP Inc.

Ruestow E G, 1996. *The Microscope in the Dutch Republic: The Shaping of Discovery*, Cambridge: CUP. Available at: http://books.google.co.uk/books?id=ESwJgm87z6MC&printsec=frontcover&source=gbs_navlinks_s#v=onepage&q=&f=false.

Russell W L, 1977. Mutation frequencies in female mice and the estimation of genetic hazards of radiation in women. *Proc N A S*, 74, pp.3523–3527.

Russell L B, 2013. The Mouse House: A brief history of the ORNL mouse-genetics program, 1947–2009. *Mutat Res*, 753, pp.69–90.

Russell W L, 1954. Genetic effects of radiation in mammals. In *Radiation Biology*. NY: McGraw-Hill; HollaenderA(ed). Available at: https://archive.org/stream/radiationbiology0102holl#page/824/mode/2up.

Russell W L, 1956. Radiation in mice:the genetic effects and their implications for man. *Bull Atomic Scientists*, 12(1), pp.19–20.

Russell W L, 1989. Reminiscences of a mouse specific-locus test addict. *Environ Mol Mutagen*, 14(Suppl 16), pp.16–22.

Russell W L, 1951. X-ray- induced mutations in mice. *Cold Spring Harbor Symp Quant Biol*, 16, pp.327–335.

Russell W L, Russell L B and Kelly E M, 1958. Radiation dose rate and mutation frequency. *Science*, 128, pp.1546–1450.

Russell W L, Russell L B, Gower J S and Maddux S C, 1958. Radiation-induced mutation rates in female mice. *Proc N A S*, 44, pp.901–905.

Sachs J von, 1890. *History of Botany (1530-1860) [English translation]*, Oxford: Clarendon. Available at: http://books.google.co.uk/books?id=1-2M9XwzpUC&pg=PA422&lpg=PA422&dq=august+henschel+plants&source=bl&ots=mgzQzsaEHp&sig=WdgwyG3vEjgEdUPRsY7f3acB8oU&hl=en&ei=qyqMSqmDEM6ZjAfHve3kCw&sa=X&oi=book_result&ct=result&resnum=2#v=onepage&q=august%20henschel%20plants&f=false.

Sankaranarayanan K, 2001. Estimation of the hereditary risks of exposure to ionizing radiation: history, current status and emerging perspectives. *Health Physics*, 80(4), pp.363–369.

Sankaranarayanan K and Chakraborty R, 2000a. Ionizing radiation and genetic risk XII The concept of potential recoverability correction factor (PRCF)and its use for predicting the risk of radiation-inducible genetic diseases in human live births. *Mutat Res*, 453, pp.129–179.

Sankaranarayanan K and Chakraborty R, 2000b. Ionizing radiation and genetic risks XI The doubling dose estimates from the mid-1950s to the present and the conceptual change to the use of human data for spontaneous mutation rates and mouse data for induced rates for doubling dose calculations. *Mutat Res*, 453, pp.107–127.

Sankaranarayanan K and Chakraborty R, 2000c. Ionizing radiation and genetic risks. XIII. Summary and synthesis of papers VI to XII and estimates of genetic risks in the year 2000. *Mutat Res*, 453, pp.183–197.

Sankaranarayanan K and Gentner N E, 2002. Reply to Comments on Hereditary effects of radiation. *J Radiol Prot*, 22(1), pp.87–92.

Sapp J, 1990. The Nine Lives of Gregor Mendel. In Le Grand H E, ed. *Experimental Enquiries*. Kluwer Acad Pub, pp. 137–166.

Sarkar S, 1992. Haldane and the emergence of theoretical population genetics, 1924-1932. *J Genet*, 71, pp.73–79.

# BIBLIOGRAPHY

Schrodinger E, 1944. *What is life?*, Cambridge: CUP.

Schull W J, 1995. *Effects of Atomic Radiation: A Half-Century of Studies from Hiroshima and Nagasaki*, New York: John Wiley.

Schwartz J, 2008. *In Pursuit of the Gene*, Cambridge MA: Harvard UP.

Searle G R, 1976. *Eugenics and Politics in Britain 1900-1914*, Leyden: Noordhoff.

Selser R and Sartwell P E, 1958. Ionizing radiation anbd longevity of physicians. *J Am Med Assoc*, 166, pp.585–587.

Shull G H, 1948. What is "heterosis"? *Genetics*, 33, pp.439–446.

Sloan P R and Fogel B ed., 2011. *Creating a Physical Biology*, Chicago: Chicago Uni Press.

Sonneborn T M, 1978. My intellectual history in relation to my contribution to science. Available at: Joshua Lederberg Papers US Nat Library Med.

Soto A M and Sonnenschein C, 2014. One hundred years of somatic mutation theory of carcinogenesis: Is it time to switch? *BioEssays*, 36(1), pp.118–120.

Springer N M and Stupar R M, 2007. Allelic variation and heterosis in maize: How do two halves make more than a whole? *Genome Res*, 17, pp.264–275.

Stadler D, 1997. Ultra-violet mutation and the chemical nature of the gene. *Genetics*, 145, pp.863–865.

Stadler L J, 1932. On the genetic nature of induced mutations in plants. In *Proceeding of the Sixth International Congress of Genetics*. Ithac, NY, pp. 274–294.

Stein G, 1937. *Everybody's Autobiography*, Random House.

Stern A M, 2005. Sterilized in the name of publiuc health. *Am J Public Health*, 95, pp.1128–1138.

Sturtevant A H, 1965. *A History of Genetics*, NY: Harper and Row. Available at: http://www.esp.org/books/sturt/history/readbook.html.

# GENES, FLIES, BOMBS...

Sturtevant A H, 2001. Reminiscences of T H Morgan. *Genetics*, 159, pp.1–5.

Sturtevant A H, 1959. T H Morgan. *Biogr Mem Nat Acd Sci*, 33, pp.283–325.

Sturtevant A H, 1925. The effects of unequal crossing over at the bar locus in Drosophila. *Genetics*, 10, pp.117–146.

Sturtevant A H, 1913. The linear arrangement of six sex-linked factors in Drosophila, as shown by their mode of association. *J Exptl Zool*, 14, pp.43–59.

Sturtevant A H, Bridges C B and Morgan T H, 1919. The spatial relations of genes. *Proc N A S*, 5(5), pp.167–173.

Sutton W S, 1902. On the morphology of the chromosome group in Brachystola magna. *Biological Bulletin*, 4, pp.24–39.

Sutton W S, 1903. The chromosomes in heredity. *Biol Bulletin*, 4, pp.231–251.

Taylor L S, 2002. Brief history of NCRP. *Health Physics*, pp.781, 776.

Taylor L S, 1990. Oral History Transcript. Available at: http://www.aip.org/history/ohilist/5153_2.html.

Taylor L S, 1984. *Tripartite conferences on radiation protection: Canada, United Kingdom, United States (1949-1953),* Wash DC:DOE OSTI / No: NVO-271.

Timofeeff-Ressovsky N , Zimmer K G and Delbruck M, 1935. Uber die Natur der Genmutation und der Genstruktur. *Nach Gesell Wissenschaften Gottingen, Math-Phys*, 6(13), pp.190–245.

Tyzzer E E, 1916. Tumor immunology. *J Cancer Res*, pp.155, 125.

UNSCEAR, 2001. *Hereditary effects of radiation*, UNSCEAR.

UNSCEAR, 1972. *Ionizing radiation:Levels and effects*, UNSCEAR.

UNSCEAR, 1962. *Report of the United Nations Scientific Committee on the Effects of Atomic Radiation.*, UNSCEAR.

# BIBLIOGRAPHY

UNSCEAR, 1958. *Report of the United Nations Scientific Committee on the Effects of Atomic Radiation. Suppl. 17 (A/3838)*, UNSCEAR.

USAEC, 1955. The effects of the hydrogen bomb. *Manchester Guardian, 1 March*.

Vavilov, Lysenko and Polyakov, 1940. Genetics in the Soviet Union: Three Speeches From the 1939 Conference on Genetics and Selection. *Science Society*, 4(3). Available at: http://www.marxists.org/subject/science/essays/speeches.htm [Accessed November 17, 2010].

Wallace B, 1991. *Fifty Years of Genetic Load: An odyssey*, Ithaca: Cornell UP.

Wallace B, 1970. *Genetic Load: Its Biological and Conceptual Aspects*, Englewood Cliffs, NJ: Prentice-Hall Inc.

Warren S, 1956. Longevity and causes of death from irradiation in physicians. *J Am Med Assoc*, 162, pp.464–468.

Warren S et al, 1949. *Minutes of the permissible doses conference held at Chalk River, Canada, September 29-20th, 1949*, / No: RM10.

Webb J H, 1949. The fogging of photographic film by radioactive contaminants in cardboard packaging materials. *Phys Rev*, 76(3), pp.375–380.

Webb S, 1910. Eugenics and the Poor Law-- The Minority report. *Eugenics Rev*, 2(3), pp.233–241.

Webber H J and others, 1914. *Eugenics - Twelve University Lectures*, NY: Dodd, Mead and Co.

Weinstein A, 1928. The production of mutations and rearrangements of genes by x-rays. *Science*, 67, pp.376–377.

Weismann A, 1889. On Heredity 1883. In *Essays upon Heredity and Kindred Biological Problems*. Oxford: Clarendon Press, pp. 67–105. Available at: http://www.esp.org/books/weismann/essays/facsimile/.

Weiss S F, The race hygiene movement in Germany 1904-1945. In Adams M B, ed. *The Wellborn Science*. NY;Oxford: OUP.

## GENES, FLIES, BOMBS...

Wells H B, 2012. *Being Lucky: Reminiscences and Reflections*, Indiana UP.

Whiting P W, 1928. The production of mutations by x-rays in Habrobracon. *Science*, 68, p.59.

Wilson C, 1995. *The Invisible World*, Princeton, NJ: Princeton UP.

Winchester S, 2002. *The Map that Changed the World*, London: Penguin Books.

Wolstenholme G ed., 1963. *Man and his Future*, Boston; Toronto: Little,Brown and Co.

Woodruff L L, 1921. History of biology. *Sci Monthly*, 12(3), pp.253–281.

# 16   INDEX

Altenburg, Edgar. 41p., 49, 53, 55p., 59, 63, 91, 231
American Breeders Association......198, 200
American Eugenics Society...............210
Amhurst College...................99, 121, 179
ankylosing spondolytis......................150
Aristotle...............................1p., 6, 22
around the world.............................218
Ascaris..........................................33pp.
Atomic Bomb Casualty Commission (ABCC)..................................118
Bardeen, C R....................................110
Bateson, William.....................30, 38pp.
Beagle...............................................26
BEAR Committee 153, 155, 158, 161, 175, 232
Becquerel, Henri...............................105
Beneden, Edouard van........................33
Bennett, Milly..............................78, 87
biometricians....................................40
Bonnet, Charles................................8p.
Bourignon, Antoinette..........................6
Boveri, Theodor............................34pp.
Brachystola.......................................36
Bragg, W H and W L........................223
Bridges, Calvin....42, 45, 47p., 50p., 65, 76, 96, 225, 233
Buck, Carrie.................................207pp.
Buffon, George....................11, 14, 23p.
Camerarius, Rudolf............................12
Carter, Toby.............................147p., 150
Casti connubii..................................214
Castle Bravo....................129p., 132, 149
Castle, W E........................................44
Chain of Being................................10pp.
Charles, D R............................136, 142p.
Chiroscope.......................................103
Ciba Foundation symposium 1962..220
Cockroft, John...........................150, 157
colloidal theory.................................223
Colwell and Russ..............................105
Congenital abnormalities..................176
Correns, Carl..................................37p.

Crick, Francis...................226, 228p.
Curie, Marie.......................106, 109
Cuvier, George.........................23pp.
Dalenpatius..................................7
Darwin, Charles...1, 11, 22, 25pp., 37, 39
Darwin's theory...........................26
Datura........................................15
Davenport, C B.....200, 202p., 205, 211, 215
Delbruck, Max...................75, 77, 226
DNA as genetic material............227pp.
DNA replication..........................229
Dominance of Economics over Eugenics ................................215
Doppler, Christian........................16
double helix..............................228
doubling dose....135, 146, 152, 156, 173, 175, 178
Drosophila melanogaster.............43p.
Edinburgh...86, 89, 92, 94, 96, 99, 112, 122, 135, 147, 218, 227
epigenesis..........................1p., 6, 8p.
Eugenics Education Society.........194
eugenics in Germany..................218
Eugenics Record Office. 202p., 205, 214
Eugenics Review, The.................195
Eugenics Society.......................195
eugenics, decline.......................213
eutelegenesis............................217
FBI........................................64, 144, 164
Films...........................................
   3000 A.D..............................165
   Attack of the Crab Monsters......163
   Behemoth the Sea Monster.......164
   On The Beach........................165
   Spiderman.............................164
   The Amazing Colossal Man......163
   The Day the Earth Caught Fire...165
   The Day the World Ended........165
   The Incredible Shrinking Man....164
   The Thing from Another World...162
   Them!....................................164
   Unknown World......................165
   War of the Colossal Beast........163
Fisher, R A................70pp., 98, 147
Fitter Families Contest................210

# GENES, FLIES, BOMBS...

Flemming, Walther..............................33
Freund, Leopold................................102
Fryer, Grace........................................108
Gaertner, Carl von........................14pp.
Galton, Francis .....................................39
genetic death.137pp., 155, 166, 173, 175, 232
Geneticists' Manifesto......................218
Geoffroy............................................24p.
germinal choice.....................220p., 231
Gray.............................................117, 133, 136
Grew, Nehemiah.................................12
Haeckel, Ernst...............................29, 33
Haldane, J B S.....71p., 74, 89, 138, 148, 173, 185, 218, 247p.
Hartsoeker, Nicklass..........................7p.
Harvey, William....................................2
Hawks, Herbert.................................103
Henschel, August.................................14
Henslow, John.....................................26
Hertwig, Oscar....................................34
Hippocrates...........................................1
Hiroshima..........................................117
His, Wilhelm........................................30
Hogben, Lancelot........................89, 218
Holmes, Oliver Wendell ..................209
homunculus...........................................7
Hooke,Robert..............................2pp., 31
Human Betterment Foundation........211
Huxley, Julian......53p., 72, 89, 98, 218, 220p.
ICRP.......................135p., 168, 171, 177p.
internal dose.......................................159
Ishmaelites.........................................198
Jansen, Sacharias..................................2
Joint Congress Hearings...................158
Jordan, David Starr........................199p.
Jukes........................197pp., 202, 249
Kallikaks............................................198
Koelreuter, Joseph...........................13pp.
Koernicke, M.....................................110
Kölliker, Albert....................................32
Lamarck, Jean-Baptiste 23pp., 28p., 31, 43
Laughlin, Harry.202p., 205, 207p., 211, 215
Leeuwenhoek, Anthony van......3p., 6, 8
Lenin....................................................79

leukaemia 105, 107, 118p., 136, 150, 152, 154, 158, 162, 167pp., 172
lifespan reduction..............................161
Linnaeus, Carl...............................12pp.
Lombardo, Cesare.............................196
Malebranche, Nicholas de....................6
Malpighi, Marcello..........................2, 4
Manhattan Project. 114pp., 126pp., 134, 142, 144p.
Martland, Harrison S........................108
martyr memorial................................104
Mavor, J W.........................................110
McClung, C E......................................35
Mechanix Illustrated.........................165
Medical Research Council. 92, 149, 158, 160
Mendel, Gregor 15pp., 22, 34, 36pp., 43, 235, 237
Miss Atomic pin-ups.........................167
modifier genes.....................................49
Montgomery, T H................................35
Morgan, Hyacinth..............................212
Morgan, Thomas Hunt.....30, 37, 43, 45
Muller, Hermann..................................
   City of Hope.................................230
   death..............................................232
   family................................................20
   health.............................................230
   House Committee on Un-American Activities..............................179
   humanist........................................231
   politics...........................179, 215p., 232
   suicide attempt................................63
multifactorial diseases....................176p.
Mutation Component.................173, 177
Nagasaki..............................................117
Nägeli, Karl....................................33p.
Nams...................................................198
Naudin, Charles...................................15
Neo-Darwinism....................................31
Neo-Lamarckism..................................29
Noyes, John..........................185pp., 222
Offermann, Carlos.................63, 78, 100
one gene-one enzyme theory............225
orthogenesis.........................................30
Osteoscope.........................................103
Out of the Night................87, 179, 216p.
Painter, Theophilus.............51, 56, 61pp.
pangenesis....................1, 28, 31, 37, 189

# INDEX

Pearson, Carl..............................40
Peithologian lecture...............42, 215pp.
Penney, William..................................118
Perthes, G............................................110
Pisum....................................................17
polonium, discovery..........................106
polytene chromosomes 51p., 56, 66, 85, 96p.
Poor Law........................................193pp.
Popenoe, Paul..............................211, 219
Potential Recovery Correction Factor ..................................................174p., 177
preformation.....................2, 6pp., 11, 14
protein structure.................................224
Pugwash conference..........................231
Purkyně, Jan..........................................31
Quackenbush, L S.................................44
Radiation Effects Research Foundation (RERF).................................................170
radiostrontium...................................160
radium..................................................
    discovery..............................106
    leukaemia.............................107
radium paint hazards........................108
radon hazards.....................................109
Raffel, Daniel.........................63, 100p.
Ray-Chaudhuri, S P......................90, 93
Réamur, Rene.....................................8p.
Remak, Robert..................................32p.
Rollins, William...................................
    x-ray precautions...............104
Röntgen, Wilhelm Conrad................102
Roux, Wilhelm.....................................34
Russell, W L........136, 143pp., 147p., 153
saltationism..........................................30
Schleiden, Matthias.........................32p.
Schrodinger, Erwin...........................226
Schwann, Theodor............................32p.
Serber, Robert....................................118
Siegesbeck, Johann..............................12
Sievert.................................................137
somatic mutations................82, 112, 154
Spallanzani, Lazzaro.........................8p.
Spark, The............................63, 100, 179
Stadler, Lewis...60p., 93, 96p., 227, 232
sterilization UK..................................212
sterilization USA.......................203, 207
Strasburger, Eduard.........................33p.
Sturtevant, A H....42p., 45, 47p., 50pp., 57, 65p., 90, 153p., 157, 159, 232p.
SUNSHINE project............................160
Sunshine Unit.....................................161
Sutton, Walter...................................35p.
Swammerdam, Jan...........................4pp.
Three Man Paper...................75, 91, 226
threshold hypothesis.........110, 112, 123, 134pp., 148p., 151, 154, 158, 168, 171, 178
Timoféeff-Ressovsky, N................39, 63
Trembley, Abraham........................8, 11
Trinity test..........................................164
Tschermak, Erich..............................37p.
Unger, Franz........................................16
UNSCEAR................168p., 172, 175, 178
US Radium Corporation...................108
Vavilov, N I..................57, 76p., 84, 87
Vestiges of the Natural History..........25
Virchow, Rudolf...............................32p.
Vries, Hugo de....................30, 37p., 43
Wallace, Alfred..............................25, 29
Watson, James...................................228
Webb, Sidney.............................194, 213
Weismann, August....................30p., 33
Weldon, W F R....................................39
What is Life?......................................226
Whitman, Charles Otis .......................43
Willis, G S...........................................107
Wilson, E B...........................................35
X-ray and Radium Protection Committee.........................................107
x-ray crystallography.......................223
Yule, George Udny.............................40

www.ingramcontent.com/pod-product-compliance
Lightning Source LLC
Chambersburg PA
CBHW070725160426
43192CB00009B/1315